THE OCCULT

This is a volume in the Arno Press collection

THE OCCULT

Advisory Editor
James Webb

See last pages of this volume
for a complete list of titles

AN

INTRODUCTION

TO

THE STUDY

OF

ANIMAL MAGNETISM

[Jean] Dupotet de Sennevoy

ARNO PRESS

A New York Times Company

New York / 1976

Editorial Supervision: ANDREA HICKS

Reprint Edition 1976 by Arno Press Inc.

Reprinted from a copy in the University
of Illinois Library

THE OCCULT
ISBN for complete set: 0-405-07936-2
See last pages of this volume for titles.

Manufactured in the United States of America

Library of Congress Cataloging in Publication Data

DuPotet de Sennevoy, Jules, baron, 1796-1881.
An introduction to the study of animal magnetism.

(The Occult)
Translation of Traité complet de magnétisme animal.
Reprint of the 1838 ed. published by Saunders & Otley, London.
1. Animal magnetism. I. Title. II. Series: The Occult (New York, 1976-)
BF1131.D85 1976 154.7'2 75-36837
ISBN 0-405-07950-8

AN

INTRODUCTION

TO

THE STUDY

OF

ANIMAL MAGNETISM.

BY

THE BARON DUPOTET DE SENNEVOY.

WITH AN APPENDIX,

CONTAINING

REPORTS OF BRITISH PRACTITIONERS IN FAVOUR OF THE SCIENCE.

——" History would be earthy were it not
For spiritual deductions; but where facts
Could be authenticated, I have ever
Given faithful record and fair inference,
Rearing a superstructure on just grounds."

COSMO DE MEDICI.

LONDON
SAUNDERS & OTLEY, CONDUIT STREET.
1838.

T. C. Savill, Printer, St. Martin's Lane, Charing Cross.

A MONSIEUR LE COMTE DE STANHOPE.

Monsieur le Comte,

Plusieurs mois s'étaient déjà écoulés en tentatives vaines pour amener les savants de votre pays à étudier les phénomènes du Magnétisme Animal ; mes nombreux appels à l'examen des faits nouveaux étaient restés sans réponse, et la presse et l'opinion que mes expériences du "North London Hospital," auraient dû éclairer, s'étaient au contraire déchainées contre moi. Ces obstacles si puissants me paraissaient insurmontables, car alors une seule voix s'était élevée pour prendre ma défense. Le cœur attristé j'allais renoncer au dessein que j'avais formé de répandre le magnétisme en Angleterre, et porter à un autre peuple la découverte si importante du Magnétisme Animal.

Vous m'encourageâtes alors, Mr. le Comte ; vous me fites espérer un changement favorable dans l'esprit de vos concitoyens, et pour la vérité dont je me suis fait le défenseur, un meilleur avenir. Confiant dans vos paroles, je restai ; et bientôt le concours de médecins distingués, et le vôtre, Mr. le Comte, vint rendre ma mission plus facile, et applanir quelques-unes des nombreuses difficultés qui s'étaient opposées à l'exécution de mes projets.

Si donc la science que je professe s'établit en Angleterre vingt années plus tôt qu'elle ne l'eût fait peut-être, si le bien qui doit en résulter pour l'humanité a déjà commencé, une partie du tribut de reconnaissance vous appartient sans doute ; et en vous dédiant cet ouvrage, j'ai voulu que ce fait ne fût pas ignoré.

Veuillez, Monsieur le Comte, recevoir l'assurance de mon profond respect,

LE BARON DUPOTET DE SENNEVOY.

20, *Wigmore-street, Cavendish-square,*
le 7 *Mai,* 1838.

CONTENTS.

APPENDIX.

PREFACE.

THE object of the present volume is to lay before the British public a compendious view of the elementary facts and principles of Animal Magnetism. Hitherto there has been in this country a disinclination to entertain this investigation; but I trust the evidence now adduced will tend to dispel this prejudice, which can only have arisen from the science not having been yet fairly represented.

It is evident that our belief in the facts of any newly-announced doctrine must be derived either from authority, or from direct personal observation. In respect to authority, the most eminent scientific men in Europe have acknowledged the facts of animal magnetism. Among others, Cuvier, La Place, Ampère, Hufeland, Treviranus, Humboldt, Sprengel, Reil, Autenreith, Brandis, Kieser, Gmelin, Georget, Cloquet, Rostan, Andral, Dugald Stewart, Coleridge, &c.; and by referring to the Appendix of this Volume it will be observed that two of the most distinguished medical professors in this city—Professors Elliotson and Mayo—have also recently convinced themselves of the reality of the magnetic influence. In addition to these authorities, I may also appeal to the Report of the Commission of the French Academy, which, after a very formal and deliberate investigation, recorded facts in support of animal magnetism, the details of which cannot, I apprehend, be in any way impugned. Hence, if reference to authority fail to command conviction in its favour, at least it should induce those who have not yet studied the science to suspend their judgment until they have investigated it.

Again: The appeal to personal observation is open to every inquirer, and can, it is obvious, be obtained only by attending the *séances* of any professor of animal magnetism, whether at the North London Hospital, or at my own apartments, where I shall always be happy to afford men of science and literature every facility for the examination of magnetical phenomena. Here, however, I would caution those who attend such demonstrations, not to form an opinion too hastily;—let them, in particular, beware how they attach importance to negative evidence. It is not pretended by magnetisers that all individuals are equally susceptible of the magnetic influence; the best magnetisers will occasionally fail to induce any ostensible effects ; furthermore, from my own experience I do not think the phenomenon of somnambulism occurs in more than three out of ten cases, yet the positive evidence is hereby not rendered the less interesting, much less is it in the slightest degree invalidated.

It remains for me to add, that this volume is intended only as an introduction to the study of animal magnetism ; it being my intention to publish, speedily, a more comprehensive and systematic work on the subject.

20, *Wigmore-street,*
Cavendish-square.

AN

INTRODUCTION TO THE STUDY

OF

ANIMAL MAGNETISM.

CHAP. I.

HISTORY OF ANIMAL MAGNETISM.

In commencing a course of demonstrations in London on Animal Magnetism, I am aware of the many difficulties that surround me, and I rely on overcoming them, only upon the truth of the cause I advocate. In Paris, when this new discovery first became the theme of discussion, it was assailed by every species of hostility; learned professors denounced it from their chairs; unlearned journalists made it the subject of their flippant pleasantries; it was turned into ridicule on the stage; and every satirist who could pen an epigram, directed the energies of his wit against it. But animal magnetism could not thus be put down. In the same city, only a few years afterwards, when the shock of political convulsions had subsided, and science and literature again resumed their dominion, the same facts again challenged attention. In vain were they denied; they were over and over again demonstrated in the public hospitals, in the presence of a numerous auditory, to the conviction of the most sceptical; and the force of the evidence alone bore down those official barriers of pre-judgment which so often obstruct the path of knowledge. Hence, the Académie Royale de Médecine was in a manner forced, by the cogent remonstrances of its own members, to appoint a new commission to investigate the subject; and after a most elaborate and scrupulous inquiry, during which every possible source of fallacy which suspicion could suggest was narrowly scrutinised, the commissioners, many of whom were previously unbelievers, not only attested the truth of animal magnetism generally, but cited, in their report, facts the most unequivocal, proving the existence of the higher order of the phenomena which its influence induces. In France, this was an important epoch in its history; but in the meantime, in Germany, Holland, Sweden, Russia, the new science spread rapidly; so that, at present, public establishments for the practice of animal magnetism exist in many of the most enlightened cities of Europe. Journals reporting its progress are periodically published; theses in its favour, by the candidates for medical degrees, are admitted and defended in the schools of medicine; and among those who, from personal observation, have attested its phenomena, may be enumerated the names of some of the most eminent scientific men of the age. On the continent, such is the present state of animal magnetism; but in England it is otherwise; here the subject has not been fairly investigated; here it is yet only imperfectly understood; and here, therefore, I expect to encounter the same obstacles, the same prejudices, even the same personal hostility which I had to contend against twenty years ago in Paris; but so firmly do I believe in the steady and immutable progression of truth, that I enter cheerfully on the task, satisfied that if the most incredulous of my antagonists will only witness and investigate the phenomena, they will become convinced of their reality.

On my arrival in this city, it was my intention to demonstrate the influence of animal magnetism before the students and medical professors in the public hospitals, and for some time, through the kindness of Dr. Elliotson, I was permitted to do so at the North London Hospital ;* but the managing committee having, in his absence, objected to my continuing my operations at this Institution, I was obliged to seek another arena ; and the apartments I have now selected are sufficiently spacious for the purpose.† This circumstance I do not mention with the intention of imputing blame to any of the parties concerned, but simply from the desire of its being clearly understood that I have never evaded confronting the most sceptical members of the medical profession, and have always courted, if temperately conducted, the most direct and rigid investigation.

Here, therefore, in the presence of many highly-distinguished visitors, I daily pursue my demonstrations—here I have convinced many of the real existence of the magnetic influence, nay, on several occasions, have magnetised the sceptics themselves,—and here I have succeeded (which is to me a more gratifying reflection) in alleviating the sufferings of many persons afflicted with chronic nervous complaints, which had resisted every ordinary method of medical treatment.

The curative effects of animal magnetism, at an early period, excited the jealousy of the medical faculty in

* Similar experiments are now conducted at the same hospital under Dr. Elliotson's own directions.

† At No. 20, Wigmore-street, Cavendish-square.

Paris; hence, the Académie Royale de Médecine, in the exercise of its royal prerogative of intolerance, issued the following law against it:—"No physician shall declare himself a partisan of animal magnetism, either by his practice or by his writings, under the penalty of being struck off the list of docteurs-régents."* This intolerant decree was dated 27th August, 1784; but it was some consolation to the professors of the new science to remember that only a few years previously, in 1745, the same august body had proscribed the practice of inoculation as "murderous," "criminal," and "magical;" that in 1636, after lengthened debates and expostulations, it prohibited the use of antimony; and shortly afterwards issued another of its royal mandates against the introduction of quinquina, or the Peruvian bark, which the sagacious ecclesiastical authorities alleged could possess no virtues, save what it derived from a compact which the Indian who had discovered it must have made with the devil. Hence it would appear, that if learned incorporate associations of science have the advantage of enjoying the collective wisdom of their members, they have also the disadvantage of being encumbered with the aggregate weight of their prejudices; and accordingly, few discoveries have ever emanated from the bosom of Royal Societies, which, in their anxiety to maintain the retrospective dignity of science, forget the intuitive progress of the human mind. It may be true that the terrors of the inquisition have been abolished, and that the stake is

* Docteur-régent was a dignity, or title, belonging to those distinguished physicians who at that time constituted the Academy of Medicine; they had the right of summoning other physicians before them, and passing judgment on their conduct.

no longer the symbol of their tyrannical power ; but such decrees as the above are calculated to perplex, harass, and torment the individuals against whom they are directed. It is clear that the above anathema of the Académie Royale de Médecine against inoculation, antimony, Peruvian bark, and animal magnetism, could no more arrest the progress of science, so far as either of these discoveries was concerned, than could the voice of Canute the Great restrain the influx of the tide before him.

The influence of corporeal apposition, either by the touch, the breath, or the saliva, in the cure of diseases, has in all ages been recognised ; but it is to Mesmer* we are indebted for the discovery of animal magnetism ; he systematised its principles—he reduced

* M. Thouret published, in 1784, a work, entitled "Recherches et Doutes sur le Magnétisme Animal," in which he exhausts much useless erudition in endeavouring to prove that the system of Mesmer was altogether borrowed from the ancients. The principal authorities he appeals to are, Paracelsus, Van Helmont, Goclenius, Burgravius, Libavius, Wirdig, Maxwel, Santanelli, Teutzel, Kircher, and Borel, from whose works he has extracted entire passages, to prove the identity of Mesmer's and their ideas. A committee of the Royal Society of Medicine was appointed to examine this work, and highly complimented the "industry," "depth," and "sagacity," of the learned author ; but, as M. Colquhoun justly observes, on the same principle, the discovery of the circulation of the blood may be referred from Harvey to Servetus, Cesalpinus, Fabricius,—the discovery of the laws of gravitation from Newton to Lucretius, Democritus, Aristotle, &c. There is scarcely a modern discovery concerning which we may not find some vague sentences in ancient authors which will bear a translation by which it may appear to have been more or less anticipated. But supposing it established that any particular series of facts, instead of being a recent discovery, had been noticed by the ancients, the circumstance of the same observations being reiterated only affords a strong presumptive argument in favour of their reality.

them to practical application. This eminent individual was born at Mersburg, in Swabia, on the 23rd of May, 1734, and studied medicine at Vienna, where he attended the lectures of Van Swieten and De Haen, and took his degree of doctor in medicine in 1766. He was a man of quick perception, vivid imagination, and persevering disposition. His inaugural dissertation, which he published and publicly defended, was on the influence of the planets on the human body,* wherein he maintained that the sun, moon, and fixed stars, mutually affect each other in their orbits; that they cause and direct on our earth a flux and reflux, not only in the sea, but in the atmosphere;

* The circumstance of Mesmer having chosen so astrological a subject as that of the influence of the planets on the human body for his inaugural dissertation, exposed him to considerable derision; but it should not be forgotten that many men of high medical reputation have not disdained the same inquiry. In 1704, the eminent Mead published his Latin treatise " On the Power and Influence of the Sun and Moon on the Human Body, and the Diseases which arise from them." In 1747, Richter published, at Göttingen, an essay "De Potestate Solis Humanum Corpus." In 1795, Balfour put forth a treatise on " Sol-lunar Influence in Fevers;" and in 1806, Murat another, " De l'Influence de la Nuit sur les Maladies." A host of authorities, ancient and modern, to the same effect, might be cited; in fact, many physicians of the present day believe that sol-lunar influence does affect the course of certain diseases, especially nervous and mental affections; and even supposing their views concerning the direct nature of this influence to be erroneous, this much is certain, that the human body, in health and in disease, is materially influenced by the vicissitudes of the atmosphere; and how far the relative positions of the heavenly bodies may disturb its equilibrium, or otherwise affect its conditions, remains yet a curious problem for the study of those interested in meteorology. In selecting, therefore, this subject for his thesis, Mesmer was not guilty of the extravagance which many, who are ignorant of the evidence which exists upon it, have attributed to him.

and affect in a similar manner all organised bodies, through the medium of a subtile and mobile fluid, which he conceived pervaded the universe, and associated all things together in mutual intercourse and harmony. About this time the mineral magnet was held in much repute as a therapeutical agent, and Father Hell, a Jesuit, and professor of astronomy at Vienna, invented steel plates of a peculiar form, which he applied with much success in the treatment of various diseases. The medicinal efficacy of these mineral plates he imagined depended on their form; but Mesmer soon discovered that although their application apparently produced certain manifest effects, yet that he could, without using them at all, by passing his hands from the head downwards towards the feet of the patient, even at a certain distance from the body, induce the same effects. He therefore came to the conclusion that the magnetic property, which he termed animal magnetism, originated in himself, and that he could at pleasure communicate it to organized or unorganized bodies. In 1773—1774, he undertook the treatment, on this principle, of a girl named Æsterline, who had been several years afflicted with severe convulsive fits; and in this case he thought he observed the flux and reflux of this fluid, and he also noticed that after every crisis she was relieved. He communicated his discovery without reserve, and his method of conducting the operation, to Baron de Stoërck, president of the faculty of medicine at Vienna, and first physician to his Majesty. His advice to Mesmer was, on no account to publish his discovery, as such an innovation in medical practice would not fail to expose him to the censure and hostility of the faculty. But notwithstanding this admonition, Mesmer adopted, in a variety of cases, the new magnetic mode of treatment; and among the number of his patients was M. Bauer, the celebrated professor of mathematics, at Vienna, who put himself under his care, and publicly attested the cure he effected.

The erroneous reports which were soon circulated abroad concerning the method he adopted, induced Mesmer to address a letter to the different scientific academies in Europe, wherein he gave an account of his discovery, and solicited their attention to the subject. The Académie des Sciences de Berlin was the only one which deigned an answer to his epistle, and in so doing passed on him the laconic sentence, " Qu'il n'était qu'un visionnaire." It appears that Mesmer himself could not understand the apathy with which his discovery was treated. " The system," said he, " which led me to the discovery of animal magnetism was not the work of a single day. Long trains of reflection had successively accumulated in my mind. Nothing but my own perseverance gave me the necessary courage to encounter the prejudices of reason and philosophy, without considering myself guilty of great temerity. The cold manner in which the first notions I ventured to publish were received, surprised me; it was altogether so unexpected. More especially did derision appear to me misplaced, proceeding as it did from the learned, and especially the medical, world, — since my system, even had it been entirely destitute of truth, would at least have been as reasonable as any of the systems which are adorned with the name of universal principles. This failure of success induced me to revise my former opinions, and so far from losing by this scrutiny, they reappeared to me clad in the brightest colours of evidence. Indeed, everything convinced me, that in addition to those principles which are admitted in the sciences, others existed which had been neglected; and I repeatedly said to myself, So long as the principles of the sciences are uncertain or false, the efforts of the greatest geniuses for the happiness and instruction of their fellow-creatures can be of no avail. I then compared medical men to travellers who have lost their way, and go further and further out of the right road, instead of turning back to find out their mistake."

In 1775, desirous of further prosecuting his researches, Mesmer travelled into Swabia and Switzerland, and at the public hospitals at Berne and Zurich, in the presence of numerous physicians, performed many remarkable cures. On his return to Vienna, he passed through Munich, where his highness the elector of Bavaria consulted him, and sought an explanation of his reported wonders. Accordingly, he performed the magnetic operation on several patients in his presence, and convinced him of the truth of the doctrine. The Académie des Sciences of Bavaria shortly afterwards elected him a member, an honour which Mesmer, in the present probationary state of the science, could not fail to have appreciated. He therefore, in 1776, made a second journey into Bavaria, and was equally successful, curing, among many other interesting cases, M. D'Osterwald, directeur de l'Académie de Munich, who had been suffering under partial amaurosis (loss of sight) and paralysis. He then returned again to Vienna, where he continued treating magnetically a variety of cases, some account of which was published in an anonymous pamphlet, entitled, "Recueil des Cures Opérées par le Magnétisme. Imprimé à Leipsic, 1778." At length, having in the most frank and unreserved manner communicated his views and mode of treatment to all in Vienna who cared to be instructed in them, and deeply interested in the success of his discovery, he determined on proceeding to Paris. His reputation had already preceded him, and on his arrival in the French capital, February, 1778, he became the object of general interest and attraction. His doctrines, however, were violently assailed; the members of the Académie Royale de Médecine summarily repudiated them, and threatened to excommunicate all who adopted or promulgated them; but in the midst of every description of abuse, satire, and contumely, he maintained a quiet self-possession, which could only have sprung from an inward conviction of the truth of the principles he maintained.

Surrounded as Mesmer was by enemies, both public and private, his unassuming manners, his manifest sincerity, his earnest yet silent enthusiasm, and above all, his benevolent disposition, conciliated for him the esteem of persons of almost all ranks and pretensions. Men of high birth, learning, and scientific eminence, crowded round him, and listened to him with delight; and many not only became proselytes to his doctrines, but set about advocating them with the most unbounded enthusiasm. Among the number of his converts was Dr. Deslon, first physician to the Comte D'Artois, and a member of the Académie Royale de Médecinê, who soon threw off the mantle of his allegiance to the tyrannous authority of the Academy, and became one of his most zealous disciples. It is impossible, indeed, to conceive the sensation which the discovery of animal magnetism excited in Paris. No theological controversy in the earlier ages of the catholic church was ever conducted with greater bitterness. On the one side, the adversaries of Mesmer, closing their eyes to the facts before them, and hurried away by an impetuous party spirit, endeavoured to proscribe the discovery as mischievous, false, and heretical; on the other side, his advocates vindicated his cause with an impatient zeal and sincerity which was equally intolerant of all doubts or explications. One party denied altogether the effects of animal magnetism, or ascribed them to the operation of some very common-place natural causes, while the other, equally indignant, exaggerated the pretensions of the new science, and carried their theories in its defence to an absurd extent. In less than eighteen months, Paris was inundated with upwards of five hundred controversial pamphlets,—erudition, sarcasm, and wit, alternately assailing and vindicating the doctrine. The prohibitory decree above cited of the Académie Royale de Médecine, against those who ventured to practise animal magnetism, was in many cases altogether disregarded, and several of the docteurs-régents who refused to tamper with their consciences, or submit to such a verdict, were deprived of the honours and emoluments of their profession. Among these were Dr. Deslon and M. Varnier, who opposed, in a very able memoir, the decision of the medical faculty against him. This memoir was drawn up by M. Fournel, and accompanied by the joint opinion of seventeen advocates, who severely and justly deprecated the arbitrary conduct of the medical faculty especially, for exacting an oath from individuals that they never should believe in magnetism, and never become disciples of the system. Upwards of thirty physicians were so denounced, and among them, M. Donglé, docteur-régent, who has given the following account of his interview with these scientific autocrats :—

" Each of the magnetising physicians received a special summons ; and almost all appeared, and were sent into a separate chamber from the judges. Each waited with impatience his summons, and walked up and down, wondering what was going to happen. At length the usher appeared, and I was first summoned, having that honour, as the eldest of the company. I entered, much surprised at not finding myself followed by any of my companions. I was requested to be seated, and the dean began by inquiring if I had paid anything for the instruction I had received in magnetism. Surprised at this question, I answered that M. Deslon did not receive money; that he only admitted medical men to observe and assist him ; that he was in the highest degree respectable, modest, and obliging ; and that of this the faculty were not ignorant. I was interrogated like a criminal, and fancied myself transported to the Great Hall of La Tournelle. They at length concluded by presenting me with a paper which I did not consider myself at liberty to subscribe. I declined signing it, and assured the faculty, as a proof of my zeal and deference, that I had not yet discovered in this method a sufficient degree of utility to induce me to make any report on it."* This inquisitorial interview between the physician, already pre-judged, and his court of accusers, was pre-

* From a pamphlet entitled, " Rapport au publique de quelques abus auxquels le Magnétisme a donné lieu. Par M. Donglé, Docteur-Régent. Paris, 1785."

cisely analogous to the arraignment of Galileo before the unrighteous judges who condemned him.*

But notwithstanding all this, the practice of animal magnetism soon prevailed to such an extent, that Louis XVI. appointed numerous scientific commissions to examine and report on the subject. One of these included Lavoisier, Bori, Bailly, Franklin, Jussieu, and many other not less illustrious individuals. The public expected much from such an areopagus; but these learned commissioners, instead of proceeding on what in this country is called the Baconian method of induction,—instead of restricting their attention to the facts which were laid before them, set about inquiring into the cause by which they were produced. They proceeded to inquire into the existence of the fluid which Mesmer had described, but it escaped their researches. They could not discover it by the sense of sight, touch, or taste; they could not collect it in masses, nor subject it to any test of measurement or weight; they therefore made a leap in the dark, and came to the conclusion, that animal magnetism did not exist, and that the facts which they had witnessed were occasioned only by imagination, imitation, and manual contact. "L'imagination," said they, "fait tout, le magnétisme animal est nul: imagination, imitation, attouchement, telles sont les vraies causes des effets attribués au magnétisme animal." †

* An anecdote is somewhere told of Galileo, that when he was condemned by the inquisitorial tribunal for demonstrating the revolution of the earth on its axis, in the presence of his judges, he stamped the ground with his foot, and exclaimed, with an air of triumph, "It moves, notwithstanding!"

† Rapport des Commissaires de la Société Royale de Médecine. Paris, 1784.

This judgment pronounced by the commissioners was attacked from all quarters. They had, it was evident, substantiated the facts of animal magnetism, and in endeavouring to overthrow the theory of Mesmer had only hazarded another infinitely more untenable. A society existed at Strasburg, which published annals, wherein immediately appeared facts of magnetisation which the explanation of the Academy could not account for; and reports to the same effect, from Nantes, Bordeaux, and Lyons, also appeared.

But that which contributed to give the system of Mesmer more stability, and which shook the respect which many might have entertained for the decision of these royal commissioners, was, that M. Jussieu dissented from his brother commissioners, refused to sign the report drawn up by Mauduyt, Audry, and Caille, and drew up a counter-report of his own, wherein he also attested the facts of animal magnetism, and only proposes another theory in explanation of them. Instead, therefore, of these commissioners settling the disputed point as to the existence or non-existence of animal magnetism; their reports only gave the subject an additional interest. Societies for the investigation and propagation of the doctrine, under the name of "Sociétés de l'Harmonie," were formed throughout France, Germany, and Switzerland, and many distinguished persons were initiated into its secrets. The Société de l'Harmonie, at Strasburg alone, was composed of one hundred and eighty members, and procès-verbaux,* or affidavits attesting the treatment and cure of various cases, were deposited

* With these procès-verbaux, sums of money were also left with the notary as a reward for any persons who could detect any fraud or deception in the statements they contained.

with public notaries, and printed copies of them distributed in all directions. In the meantime, Mesmer, for the benefit of his health, withdrew to Spa, whither he was followed by numerous persons of rank and fortune, who were still desirous of placing themselves under his treatment. I regret I have not space to enter into further details concerning his personal history; suffice it to add, that, after receiving a subscription, which was raised in compensation for his having devoted his life to the promulgation of so important a discovery, he retired to his native place, where he died, much beloved and honoured, on the 5th of March, 1815, at the advanced age of eighty-one.

While animal magnetism was thus making a steady and rapid progress in France, political events, which shook the foundations of society, absorbed every consideration. The French revolution broke out, and threw a temporary cloud over science, literature, and the arts. The Sociétés de l'Harmonie were many of them dissolved, and the patrons and disciples of Mesmer, for personal safety, expatriated themselves, carrying with them the discovery of animal magnetism back to its native land, and diffusing its principles through Germany and Holland, Sweden and Russia. Some, too, of the pupils of Mesmer carried the doctrine as far as America: thus do all great truths, in spite of every opposition, gradually become diffused over the face of the civilized globe. When the terrors of the French revolution had subsided, and society in some measure recovered its equilibrium, animal magnetism participated in the advancement of the other sciences. In almost all the university towns of Germany, public lectures on the subject were given; the Académie des

Sciences de Berlin, which had denounced Mesmer as a visionary, now offered a premium of 3,300 francs—750 dollars—for the best explanatory thesis on the science; and a public hospital, in that city, for the treatment of patients by animal magnetism, was opened.*

In Prussia, Austria, Bavaria, Russia, laws were passed by the respective governments, prohibiting its practice by all persons, excepting by physicians regularly educated. In France, the discovery by the Marquis de Puységur of magnetic somnambulism had already thrown an additional charm over its study, and enriched physiology and psychology with a newly-developed class of facts in the highest degree curious and interesting. Next to Puységur appeared the Abbé Faria, who opened in Paris an institution for magnetism which attracted the attention of many scientific men. He was himself endued with the magnetic power to a degree which would have appeared incredible, were it not a notorious fact that its intensity augments with exercise. The daily increasing converts to magnetism attached themselves to this man of wonders; they listened with interest to his lectures, and learned by his instructions how themselves to conduct the operation. It is impossible to conceive a more direct and conclusive *experimentum crucis* than this. The mag-

* I am sensible that I ought not to pass over in silence the names of Wienholt, Gmelin, Scherf, Böckman, Kluge, Ennemoser, Lichtenstad, Stieglitz, Wolfahrt, and many other eminent men, to whom the science has been, in Germany, indebted; but in this little volume I can only give a very brief outline of its history, besides which, I wish to induce conviction by appealing to present facts,—the phenomena I daily demonstrate,—rather than by reference to any retrospective or collateral evidence, however satisfactory or conclusive.

netiser assumed not to himself any exclusive privilege, nor did he pretend to be initiated in any mysterious or incommunicable cabalistic art; on the contrary, he assured his auditors that they had only to investigate the phenomena to be convinced of their reality, and invited them in their turn to try the operation themselves, assuring them that they would be as successful as himself, if they only complied with the elementary principles of the science, which are in themselves simple and explicit, and equally intelligible to all. The instructions of the Abbé Faria were by many followed,—his predictions verified: and who that has thus practically studied animal magnetism has ever turned away sceptical from it?

The foundation of the science having been thus established on evidence the most indisputable, three separate schools of magnetism arose: — 1st, the original school of Mesmer. This prevailed principally in Paris. Its doctrines were very similar to those of the Epicurean philosophy, as described in the poem of Lucretius. Its disciples believed in the existence of the universal fluid above described, and conducted the operation physically,—that is, by passing the hands immediately over, or at a short distance from, the body of the patient. 2nd, the school of the Chevalier de Barbarin. This was founded at Lyons, and, although it had many partisans in France, prevailed principally in Sweden and Germany. Its principles remind us of the Platonic philosophy; its disciples maintained that the magnetic operation depended entirely upon a pure "effort of the soul," and was to be conducted only upon psychical principles. They were therefore termed spiritualists. Lastly, the school of the Marquis de Puységur, founded at Strasburg, the disciples of which, professing to be guided only by observation, called themselves experimentalists. The characteristic feature of this school is, that it combines the physical treatment of the school of Mesmer with the psychical treatment of that of Barbarin. That which I consider the best method of conducting the magnetical operation will hereafter be described; but in the meantime I may observe, that notwithstanding the magnetisers divided themselves into these different sects, they all maintained the same fundamental principles: they differed in theory, but each school agreed in produçing the same practical results.

While animal magnetism was thus spreading rapidly in all directions, numerous detached pamphlets and treatises appeared in its defence; but it was obvious that a good elementary, systematic work on the subject was still wanted. This *desideratum* was not supplied until 1813, when Deleuze* published his "Histoire Critique du Magnétisme Animal." The Author, who held the situation of librarian, and enjoyed the rank of Professor of Natural History at the Jardin des Plantes, in Paris, was highly esteemed in all classes of society. His scientific knowledge was well known; his moral integrity unimpeachable; the book, too, was written in a tone of candour and moderation which carried conviction along with it. No one ventured to challenge the truth of the facts which he described; but he was forced to pay the penalty of his veracity, inasmuch as, on account of his having published this critical history, he was refused admitance as a member

* In the last number of Silliman's American Journal, [October, 1837,] a translation of Deleuze's "Histoire Critique du Magnétisme," is announced, by Thomas E. Hartshorn, of Providence, Rhode Island, "with notes, referring to cases in America."

of the Institute, although he had every qualification which entitled him to the honour. It is really melancholy—nay, it is difficult to suppress our indignation—when we see the generous advocates of truth thus assailed by every species of paltry persecution. Ye men of science and literature, who boast of the advancement of the human intellect, and arrogate to yourselves the superior advantage of living in the nineteenth century, be not deceived; but examine well whether the same desolating spirit of intolerance be not yet abroad, which so long obstructed the progress of science, and cast a withering blight over the fields of knowledge.

But to return. After the appearance of the Histoire Critique du Magnétisme, by Deleuze, many men of the highest scientific eminence felt compelled, by the cogency of the evidence in its favour, to examine the facts; and the opinions expressed in favour of animal magnetism by Ampère, Laplace, Cuvier, and other men of equal eminence, any one of whose names ought to have silenced the voice of factious scepticism, induced many who were previously indifferent to investigate the subject. This was in 1819; I was then young, and a medical student, a stranger to all the prejudices of another age. I was yet ignorant that physical facts were to be rejected because they were inexplicable, or because they were contrary to received doctrines. On animal magnetism I had already thought and read much; and, never doubting but that truth would eventually be triumphant, I solicited permission to make experiments at the Hôtel-Dieu. My request was granted, and the results published in a little brochure, entitled "Exposé des Expériences de l'Hôtel-Dieu," which was translated into German and Italian, and passed through three French editions. It may be sufficient to state here that my success was complete. Among other cases which I treated was that of a girl who had been thirteen months in the hospital, suffering from hœmatemesis, (vomiting of blood from the stomach,) which had resisted the ordinary methods of medical treatment. In the presence of forty physicians I magnetised this patient, who received such decided and manifest benefit from the operation, that in twenty-seven days she left the Hôtel-Dieu perfectly cured. In this case, the vomiting of blood, which had immediately subsided on the first magnetisation, returned on its being suspended; and on the second operation again ceased, and did not afterwards recur. It may be added, that the magnetic treatment induced in this patient all the extraordinary phenomena of somnambulism; and every trial which incredulity and ingenuity could suggest was had recourse to with the view of ascertaining whether there could be any possible collusion. An account of the phenomena, as they had presented themselves, was afterwards drawn up and signed by MM. Husson, Geoffroy, Récamier, Bricheteau, Delens, &c., in fact, by all the medical men who were present, and the document was deposited with a notary in Paris. At the Hospice de la Pitié I afterwards performed other magnetic operations, which led to still further trials; after which, MM. Georget and Rostan, both eminent pathologists, published their observations,—the former in his work on the physiology of the brain, the latter in the article *Magnétisme*, published in a medical dictionary, in 1823. Both were prepared for the consequences of their avowal in favour of animal magnetism. They knew that the spirit of the age would be strongly opposed to their assertions, but confided in futurity for the justification of their opinions.

At the Salpétrière, also, M. Esquirol authorized experiments on the epileptic patients, of whom there were many in the hospital; and there, also, the magnetic phenomena were produced in the midst of a crowd of medical men, all anxious to verify, personally, facts of so extraordinary a nature. At the Bicêtre a similar scientific investigation was conducted; but it was at the Hospice de la Charité that animal magnetism was most extensively pursued. Here, in the presence of about three hundred medical men, experiments were conducted by Dr. Foissac, which were demonstratively conclusive as to the production of all the higher order of the phenomena. Nearly at the same time other experiments were made at Val-de-Grace. I magnetised there, in the presence of many students and medical men, an epileptic soldier, who experienced very marked effects. A young physician attached to the hospital, M. Desruelles, submitted himself to magnetisation, and the action was so violent that after a few minutes I was obliged to desist operating. Other experiments, equally convincing, were performed in the presence of M. Broussais, and seven or eight other physicians. Dr. Frappart threw into somnambulism a soldier affected with a nervous complaint; and the trials made upon him left no doubt respecting the power of magnetism. So convinced was M. Broussais on the subject, that he authorized the publication of the facts.

A distinguished young physician, Dr. Bertrand, whose premature death was an irreparable loss to science, established, about this period, conferences on magnetism at his own residence, which were attended by many scientific men. Many members of the chamber of deputies also interested themselves in the progress of the science. M. Chardel, the Comte de Gestas, M. de Lascases, &c., opened their saloons to those who were desirous of being instructed in magnetism; and the *corps medical* of Paris were in a manner compelled to give the subject more serious attention. The report of Lavoisier, Franklin, Bailly, &c., in 1784, was no more appealed to as the death warrant of the system; it was no longer alleged that there was no truth in Animal Magnetism because these learned academicians had ascribed all its effects to imagination, imitation, contact, &c. The constant accumulation of facts which were incontrovertible succeeded in overturning so effectually this long-disputed academical verdict, that one of the members of the Academy felt it his duty to propose that a new special commission should be appointed to investigate the subject. This was a perplexing proposition; the Academy knew not which side to take. It was composed of many members who were convinced of the truth of animal magnetism, and many who were undecided, and desirous of receiving further information. The older members were unwilling that the magnetisers should derive any importance from their decision; they were even apprehensive that the bare nomination of a committee of inquiry would lead to an impression that they were favourably disposed towards the doctrine; and several were naturally enough disinclined to disturb, much less to rescind, the decision against it contained in the old report. But the case was urgent; they therefore adopted an intermediate course. They appointed a committee, MM. Adelon, Burdin, Marc, Pariset, and Husson, to make a preliminary investigation, and report whether the Academy, without compromising itself, could appoint a new commission. This was surely proceeding with all due precaution,—and what was the result? The committee convinced itself that the evidence was sufficiently strong in favour of animal magnetism to warrant the Academy in authorising another official investigation; and on the 13th of December, 1825, some months after their appointment, the commissioners, faithful to their duty, presented themselves to deliver their report. There was on that day an excitement among the members in the council room which did not terminate with the sitting. Three days were occupied by the debate, when at length the Academy decided, by a majority of ten, in favour of the appointment of a new commission. Eleven members were therefore nominated—viz., Bourdois de la Motte (the President), Fouquier, Gueneau de Mussy, Guersent, Husson, Itard, Marc, J. J. Leroux, Thillay, Double, and Magendie; and they immediately began their labours by an address to all magnetisers, inviting them to exhibit, in their presence, the effects they professed to induce; and some hospitals even were placed at their disposal. Very few replied to this amicable appeal, because this inquiry gave publicity to those who took part in it; and many who believed in magnetism, and practised it, were not willing to come forward, as a failure would have exposed them to ridicule.

The chair of the Faculty of Medicine of Paris now resounded for the first time with the language of magnetism. Professor Rostan pronounced an apologetic discourse on Mesmerism, and described its wonderful effects, in the presence of more than four hundred of his pupils.

Lhullier Vinslow, in a treatise on Materia Medica, acknowledged the existence of magnetism, without any restriction. Professor Segalas, in his private lectures, also undertook its defence, frankly acknowledging that he had been incredulous, but that, having magnetised, he had produced extraordinary phenomena. Marjorlin recommended its employment in certain nervous complaints; Fouquier did the same; and magnetising physicians were now seen for the first time since 1784. At this period it occurred to me to give public lectures on the subject, and more than eight hundred young men received instructions from me—very imperfect, it may be, but nevertheless, they learned what was meant by animal magnetism.

But to return to the proceedings of the newly-appointed commission. Five years having elapsed, and it not having yet accomplished its task, some anxiety was felt as to the motives for so long a delay, and fears were entertained lest there should be any evasion of their duty on the part of the commissioners. Several magnetisers had presented facts to their notice, and I had also placed myself at their disposal. They had met at my residence several times to witness experiments, and I had received their promise that the effects produced before them should be acknowledged. At length, the commissioners announced that they would read their report to the Academy on the 20th of June, 1831. The hall in which the Academy assembled, so empty on ordinary occasions, was crowded that day; even the passages were obstructed by the curious. It might have been supposed that one of those decrees on which the weal or woe of a nation depends was in agitation. All the members of the Academy, even those enfeebled by age, were at their posts. The meeting was then opened, and M. Husson, the reporter of the committee, appeared at the bar with a voluminous roll of paper in his hand. He spoke in a grave and somewhat measured tone, and began by reviewing the events which had preceded the nomination of the commission; he then invoked the memory of the ancient friendship borne by the elder practitioners present, to its members;—he at length came to the point, and related, first, all that had appeared doubtful to the commissioners; then he cited the facts which could be explained by causes foreign to magnetism; and lastly, the positive facts which could only be referred to magnetism itself. But as the forms of oratory, and the cases already cited, had absorbed much time, the conclusion of the report was deferred until the next meeting.

This was the day of the grand battle;—I say, battle, for on this occasion there was a general affray among the members of the Academy. While the report was being read, they listened with uneasiness to the facts detailed; but when the higher phenomena of lucidity were described, a general murmur, not very flattering to the commissioners, prevailed, which gradually increased, until several Academicians started from their seats, and apostrophised, in unmeasured terms of indignation and contumely, the men who had conscientiously related the facts which they had seen and attested. An outcry was raised on all sides against the commissioners; but, without being disconcerted, many members of the Academy, who believed in magnetism because they had themselves examined it, vindicated them, and retorted: — "You do not believe in magnetism — be it so; but in this very place the circulation of the blood was denied; yet the blood does circulate! Here they who practised inoculation were denounced as impostors, and the inoculated as dupes and idiots; yet was inoculation no imposition or idiotcy;—here, also, the physicians who first employed tartar emetic were put on their trial, and expelled the Academy; yet we have now for colleagues men who employ it in enormous doses!"* Another exclaimed, with stentorian voice, "This Institution ridiculed those who affirmed they had seen stones fall from the sky; yet meteoric stones do fall!" Thus the sanctuary of science was on a sudden converted into an arena of Babel-like confusion, because a few of the learned members of the Academy were not prepared to accredit the facts which the commissioners, whom they had themselves delegated with the inquiry, reported to them. But it was necessary to terminate so turbulent a discussion; and then came the question whether the report should be published or not, which, after another stormy contention, was decided in the negative. As most of the members, however, wished a copy of it for themselves, it was agreed that a limited number of copies should be lithographed.†

* At that period Récamier administered tartar emetic in doses of twenty, forty, eighty, and one hundred and sixty grains.

† This report, in 1833, was translated into English by M. Colquhoun,—an advocate at the Scottish bar,—to whom animal magnetism in this country is much indebted. He afterwards, in 1836, published the "Isis Revelata," in two volumes, which presents us with an admirable account of the origin, progress, and present state of the science. This latter work contains a mass of interesting information on the subject; besides which, it is written in a philosophical tone, and pervaded throughout by a spirit of persuasive eloquence which is as creditable to the feelings as to the judgment of the learned author. It is impossible to conceive any impartial mind rising sceptical after its perusal.

The Académie Royale de Médecine thus put on record clear and authenticated evidence in favour of animal magnetism. We have first, be it observed, the favourable opinion, founded on experimental inquiry, of the preliminary committee, appointed to investigate whether the evidence were sufficiently strong to warrant the Academy in appointing a new commission; we have next the report of the new commission, detailing circumstantially the facts which its members witnessed, and the methods they adopted to detect every possible source of deception. It should also be remembered that many of the commissioners, when they entered on the investigation, were not only unfavourable to magnetism, but avowedly unbelievers,* so that their evidence in any court of justice would be esteemed the most unexceptionable that could possibly be desired. They were inquiring, too, not into any speculative or occult theory, upon which there might reasonably be a chance of their being led away by sophistical representations; but they were inquiring into the existence of facts only—plain, demonstrable facts—which were in their own nature palpable to every observer. Here there was no ground for evasion; but I do not seek to convince any persons in this country by appealing to the evidence, satisfactory as it may be, contained in this report, nor do I solicit their belief on the faith of any existing scientific authorities, however high or estimable,—I invite the men of science and letters in this enlightened city to witness

* One of the incredulous members, M. Itard, subjected himself, as will be hereafter seen, to the magnetic action. I operated. The effect was decisive, and convinced all who were present of the reality of the magnetic influence. What further or more conclusive evidence could be desired?

the phenomena themselves, and critically examine them. If animal magnetism be a fraud or a delusion, let them openly unmask the imposition; if, on the contrary, its principles be, as I believe, founded on truth, let them not shrink back on the threshold of a science which opens a new field for investigation, and which promises to confer the most signal blessings on suffering humanity.

CHAPTER II.

PHYSICAL PHENOMENA OF ANIMAL MAGNETISM.

THE effects of animal magnetism have been described as marvellous, incredible, contrary to the course of nature: this is exaggeration. It is true that many of them are very remarkable, but they are not more extraordinary, nor are they more inexplicable, than many other physiological and psychological phenomena which frequently occur in normal and abnormal conditions of the human body. He who would study the philosophy of organisation must expect to meet with marvels; his path will be infallibly beset by mystery. " The greatest wonder of all wonders," observed Lessing, " is, that we are insensible of the wonders that daily surround us." The phenomena of animal magnetism are extremely various, the reason of which may be satisfactorily explained. Thus, in the physical sciences, chemistry, electricity, galvanism, the conditions being always the same under which certain experiments are performed, the effects may with tolerable certainty be predicated; but in animal magnetism this is not the case, because the conditions under which the operation is performed can never be precisely similar. The power of the magnetiser, the energy of his volition, and the sensibility or constitution of the person operated upon, must in every case vary; and hence the effect produced must be very much diversified; yet an intelligent eye will not fail to perceive that these varied effects are in reality modifications of the same phenomena, resulting from the operation of the same cause, the laws of which may, it is true, be at present only imperfectly understood; but I am nevertheless persuaded that animal magnetism—that active principle which we possess within us, and which, under the energy of our volition, manifests itself by the effects it visibly induces—does possess fixed and constant laws, which the progress of knowledge will eventually reveal to us. Let us not, therefore, in the meantime, indulge in hypothetical conjectures, and fly, as fancy may suggest, from one uncertain theory to another; but let us restrict our attention to facts which are attested by men who are worthy of credit and competent observers—facts which I profess myself capable of reproducing whenever the occasion may require it.

The variety of the phenomena which are produced, as above stated, by the magnetic operation, renders it difficult to classify them. I shall, however, adopt the course usually observed in explaining the physical sciences. I shall proceed gradually from the effects which are most simple to those which are more and more complex; but even this can only be done in an arbitrary manner, because the higher phenomena are often developed in conjunction with those which are much less so; and on this account, many persons in the commencement of the study are startled, and their incredulity is excited by the appearance of facts which they are not prepared to receive. Such, too, is the disposition of the human mind to theorise, that many set about explaining facts before they are assured of their existence, or can appreciate their specific character, so that many of the higher magnetic phenomena seem to carry such persons back at once into those remote fabulous ages when every unusual occurrence was pregnant with mystery, and conjured up phantoms of superstition, which, when the mind is darkened by ignorance, constantly impose on human credulity. But it is well known, that as science advances the light of truth gradually dispels these imaginary prodigies, and those appearances which were once deemed supernatural, lose entirely the character of the marvellous, and become associated with the ordinary phenomena of daily observation. Let those, therefore, who are about to enter on this investigation, not prematurely exclaim, "Extraordinary!" "Impossible!" I shall cite no fact that is not well, nay, incontrovertibly established; not one that I am not myself prepared to demonstrate.

When an individual subjects himself to the magnetic operation, the change which is thereby produced in his habitual mode of being, is often very sudden and manifest; but more frequently it is necessary for the magnetiser to persevere for awhile before any ostensible effect is induced. This will depend, it is obvious, on the power of the magnetiser and the susceptibility of the person magnetised; but although in some cases no effect be externally manifested, it will in the sequel be found that no individual can magnetise another without producing some change in his organisation; nay, sometimes the effect of the magnetic action is not felt until some time after the operation. The symptoms most commonly induced are the following :—slight pricking and winking of the eyelids—an increase, or perhaps diminution, in the pulsations of the heart—a sensible alteration in the temperature of the body—the cheeks sometimes are flushed, or become extremely pale—the expression of the countenance, indeed, undergoes a remarkable change—stretchings of the limbs and deep yawnings succeed—a gurgling noise (borborygmus) is often heard in the throat—the patient is, perhaps, disposed to move, yet feels unable to do so, or experiences an unusual sense of composure, which is to him a peculiar, an undefined delight—the breathing frequently becomes much affected, and by a singular anomaly, occasionally the circulation increases in rapidity, while the respiratory movements of the chest become less and less frequent. In one case, particularly, which fell under my observation, the pulse previous to the operation was sixty-five, the inspirations twenty-four per minute ; after the operation, however, the pulse rose to one hundred and fifteen, and one hundred and twenty, while the inspirations fell to twelve. These are the primary and most simple effects of animal magnetism ; but often, under circumstances which it is previously impossible to determine, phenomena of a more remarkable character are developed. The eyelids of the magnetisee appear spasmodically affected, and close against his will ; in vain does he attempt to open them, or change his attitude, in order to keep himself awake ; for if the magnetiser persevere, he yields gradually to his influence, and sleep, more or less profound, supervenes. His head, by its own weight, inclines forward upon the chest, or more rarely, is thrown backwards ; his eye-lids are generally half open, and the eye-ball moves slowly in the socket ; its motions may be followed by the observer, who will perceive it gradually become fixed; drops of mucus fall from the lips, the limbs become cold, and the respiration audible. If spoken to, the magnetic sleeper may perhaps attempt an answer, and appear manifestly unable to speak, or he will suddenly awake, rub his eyes, stare round him with astonishment, and recollect what has passed, as we may recal a dream. To disturb any one in this state is highly improper, for convulsive fits may be thereby occasioned; indeed, the curiosity of the idle, and their wanton interference with persons in a magnetic state, may, when they are least aware of it, lead to dangerous consequences.

In many instances the magnetic action induces much agitation, and partial or general convulsions, phenomena, which are so striking that they naturally attracted the principal attention of Mesmer and his immediate disciples. He termed these convulsions crises, because he presumed that during them the disease, thus magnetically treated, underwent a specific alleviation or cure. That these remarkable conditions are induced by animal magnetism, I prove daily,—but, lest my authority should be impugned as *ex parte* evidence, I shall here adduce testimony which may defy the suspicion of the most sceptical, inasmuch as it is the evidence of the commissioners who, in 1784, repudiated the theory of Mesmer; men who were notoriously so hostile to the science that they grasped at every shadow which could throw a doubt upon it.

It is from their memorable report, which was, some years ago, so often cited against animal magnetism, that I extract the following graphic account of the effects which its influence often induces. "The patients," observe the commissioners, "present a spectacle extremely varied, in proportion to their different habits of body; some of them are calm, tranquil, and unconscious to any sensation; others cough, expectorate, are affected with a slight degree of pain, a partial or general heat, and increased perspiration; others are agitated and tormented with convulsions. These convulsions are of extraordinary violence and duration; as soon as one person is convulsed, several others are affected by the same symptoms. The commissioners saw crises of this kind which lasted upwards of three hours, and which were accompanied by a thick and viscid fluid brought away by the violence of the effort. * * * Nothing can be more astonishing than these convulsions; he that has not witnessed the scene can have no idea of it, and on beholding it, the spectator is no less struck with the profound repose of one class of patients than with the violence which agitates another. He observes with admiration the various incidents that are repeated, and the sympathies that are developed. He sees some patients seek each other with eagerness, and in approaching, smile, converse with all the demonstrations of attachment, and soothe their mutual crises. They are entirely under the government of the person who distributes the magnetic virtue. In vain may they appear to be in a state of extreme drowsiness, his voice, his look, a motion made by him, arouses them. It is impossible not to recognise in these uniform and constant effects an extraordinary influence acting upon the patients, making itself master of them; and the magnetiser, he who superintends the process, appears to be the common centre whence this extraordinary power diverges."*

* Rapport des Commissaires chargés par le Roi de l'examen du Magnétisme Animal. *Imprimé par ordre du Roi*, à l'imprimerie Royale. Paris, 1784. This Report had a very wide circulation; upwards of 20,000 copies were issued.

This, be it remembered, is the concurrent testimony of Lavoisier, Bori, Bailly, Franklin, &c.; but it is superfluous to dwell on facts which were established in a former age; I revert to them only because I am anxious that the reality of their existence shall be placed beyond the possibility of a doubt.

It has been already stated that the magnetic operation induces sleep; it is a state of sleep, however, very different from natural sleep; its phenomena were first described by the Marquis de Puységur, but, in conformity with the plan already laid down, I shall detail its physical conditions before proceeding to those phenomena, which are of a purely psychical nature, and in the highest degree interesting, inasmuch as they reveal to us the manifestations of a spiritual existence, independent of the ordinary instruments of corporeal organisation. In contradistinction to the natural, this has been termed the magnetic sleep, or somnambulism, which may be more or less complete,—that is to say, its phenomema may be more or less perfectly developed. In this peculiar state of sleep, the surface of the body is sometimes acutely sensible,—but more frequently the sense of feeling is absolutely annihilated. The jaws are firmly locked, and resist every effort to wrench them open; the joints are often rigid, and the limbs inflexible; and not only is the sense of feeling, but the senses of smell, hearing, and sight, also, are so deadened to all external impressions, that no pungent odour, loud report, or glare of light, can excite them in the slightest degree. The body may be pricked, pinched, lacerated, or burnt; fumes of concentrated liquid ammonia may be passed up the nostrils; the loudest reports suddenly made close upon the ear; dazzling and intense light may be thrown upon the pupil of the eye; yet so profound is the physical state of lethargy, that the sleeper will remain undisturbed, and insensible to tortures, which, in the waking state, would be intolerable. These may appear startling assertions, and it should be conceded, that in proportion as any alleged facts may appear extraordinary, do they demand additional weight of evidence to establish them. This state of insensibility, now described, therefore requires to be proved by the most direct positive evidence; it must be evidence, at once clear, unequivocal, and above all suspicion; it must be evidence, too, that is tangible to those who will entertain the investigation, and this I am prepared to adduce. Can more in reason be desired? I cite authorities, I refer to scenes which occurred in public hospitals, in the presence of living witnesses, conscientious and competent observers, only because I cannot bring the experiments under the eye of the reader while perusing this page; but if he choose to visit any of my public *séances*, he may convince himself, by personal observation, of the truth of the facts which are here stated; and I make this appeal, fully sensible how difficult it is to disturb that apathy which scepticism usually engenders, or to excite even the desire of inquiry in minds which have already come prematurely to a conclusion.

Accordingly, I now proceed with the evidence. In 1820, numerous experiments were performed by me in Paris, at the Hôtel-Dieu, and many incredulous physicians, attracted by the novelty, witnessed them, and wished to satisfy themselves that there was no deception. To this I assented, and accordingly they proceeded to prove the insensibility of the patients by a variety of tests, many of which were very cruel; but these experiments were not performed by me; it was a sort of sacrifice of humanity which incredulity insisted upon, and in the infancy of the science I had no alternative excepting to permit them, or allow truth itself to be compromised by the refusal. Hence the unconscious patients were subjected to the following trials:—their lips and nostrils were tickled with feathers; their skin was pinched until ecchymoses (bruises) were produced; smoke was introduced into the nasal passages, and the feet of one female were plunged into a strong infusion of mustard-seed at a high temperature. But not the slightest sign of pain did they evince. The expression of the countenance remained unchanged, nor was the pulse in any degree affected. On being awakened, however, out of the magnetic sleep, they all experienced the pain usually attendant on such applications, and were exceedingly angry at the treatment they had received. Many of the physicians who had witnessed this scene, and who were convinced of the reality of the magnetic influence, applied to me to teach them how to conduct the operation, and several very soon acquired the method, upon which they proceeded to convince themselves, by their own personal experiments, of the absolute extinction of sensibility in such cases. The means they adopted were on many occasions revolting, but the result removed every shade of doubt from their minds. I now, therefore, proceed to adduce their evidence. It is no longer I who attest the facts; and praise or blame, should either be bestowed, will be alike inapplicable to me. Here, therefore, I subjoin the following *procès-verbal* by Dr. Robouam, who was then attached to the Hôtel-Dieu.

"I, the undersigned, certify, that on the 6th of

January, 1821, M. Récamier, on visiting the hospital, requested me to put into magnetic sleep a man named Starin, whose bed was No. 8 of the ward Sainte-Madelaine. M. Récamier first threatened him with the application of moxa,* if he allowed himself to fall asleep, and I caused the patient, much against his will, to pass into the magnetic sleep, during which M. Récamier applied the moxa on the fore-part of the right thigh, which moxa produced an eschar seventeen lines in length and eleven in breadth; that Starin did not manifest the least sign of pain, either by crying out, gesture, or variation of the pulse; that he did not feel the application of the moxa until I had roused him out of the magnetic sleep. Madame Sainte-Monique, the matron of the ward, MM. Gibert, La Peyre, Bergeret, Carquet, Truche, &c., &c., were present during this *séance*." Here, also, is another *procès-verbal* by the same physician. "I, the undersigned, certify, that on the 8th of January, 1821, at the request of M. Récamier, I put into magnetic sleep the woman Leroy (Lise), lying in the bed No. 22 in the ward Sainte-Agnès. M. Récamier had previously threatened that he would apply a moxa to her person if she allowed herself to fall asleep. I, Robouam, then, much against her will, caused the patient to fall into magnetic sleep, during which M. Gibert burned agaric under her nostrils, and this nauseous smell pro-

* The unprofessional reader should be apprised that by moxa is meant the direct application of fire to the body, as a counter-irritant. The down of the mugwort (*artimisia chinensis*), hemp, flax, any combustible substance, being rolled into a cylindrical form, is lighted, and the point of the flame, by means of a small canula, or blow-pipe, directed on the part.

duced no perceptible effect; that afterwards M. Récamier himself applied moxa on the epigastric region (pit of the stomach), which produced an eschar of fifteen lines in length and nine in breadth; that during the operation, the patient did not shew the least symptoms of suffering, either by cries, gestures, or variation of the pulse; that she remained in a state of perfect insensibility, and that on awakening from the magnetic sleep, she shewed signs of suffering great pain. MM. Gibert, Créqui, &c., were present at this *séance*." (Signed,) Robouam, Docteur-Médecin.

The next *procès-verbal* I shall cite was deposited at the office of M. Dubois, a well-known notary, at Paris, and refers to a case which was witnessed by MM. Husson, Bricheteau, Delens, and other eminent physicians. But before proceeding, it may be necessary so far to anticipate the psychical details as to observe that the individual in this somnambulic state will often exclaim aloud, converse, sing, and perform a variety of actions, as persons addicted to spontaneous or natural somnambulism often do, only in a much more perfect or lucid manner. There is also a peculiar relation established between the magnetiser and the magnetisee; they are said to be *en rapport* with each other; in other words, the magnetic sleeper, while absolutely insensible to all other external impressions, is mentally conscious of every act of the magnetiser; he will hear, even at a distance, the slightest modulation of his voice, however inaudible it may be to those around; but the power of the magnetiser over the magnetisee will be hereafter more fully described. To return, therefore, to the *procès-verbal* now cited: "Catherine Samson (a young girl of much natural timidity) was put to sleep in about fifteen minutes. Many persons present endeavoured to rouse her by individually, and then altogether, screaming suddenly in her ears. They also struck violently with their clenched fists various pieces of furniture, but could not obtain any symptom of her hearing the loudest noise. 7th September, 1820. She fell asleep in three minutes; M. Récamier opened her eyelids, shook her violently, grasped her hands violently, struck the table with all his might, pinched her repeatedly, raised her from her seat and suddenly let her fall,—still no change was perceptible, nothing which could convey an idea that the patient either heard or felt. When the magnetiser, however, spoke, she heard him distinctly. M. Récamier then alternated his voice with that of the magnetiser, but to his voice she was insensible. 10th of November. The patient fell asleep in one minute. Her lips and nostrils were tickled with feathers, yet she felt nothing. They cried out that she was an impostor, that her conduct was scandalous, and that they would kick her out of the room. Some counterfeited the voice of the magnetiser, but could obtain no answer, nor was any alteration visible in the expression of her countenance. 10th November; evening. She was magnetised in her bed; in a few minutes fell asleep, and remained all night in a somnambulic state. The persons appointed to watch her observed that she never moved; they pulled and plucked out the hair of her head by the roots, but could detect no sign of sensation."

On the 29th December, 1826, in the presence of the commissioners appointed by the Academy of Medicine, I was again called upon to magnetise this subject, and her case now attracted considerable attention, inasmuch as during the discussions which took place in the Academy, in consequence of the proposal to submit magnetism to a new investigation, M. Récamier alleged that this girl, whom the magnetisers had professed to have cured, had applied for re-admission into the hospital, where she died. This was not the fact. Six years after her pretended death, she re-appeared, and from the report drawn up by the commissioners, I subjoin the following particulars:—

"We were first of all desirous of ascertaining whether the individual presented to us by M. Dupotet, whose good faith was perfectly well known to us, was the identical person who, nine years before, had been magnetised at the Hôtel-Dieu. MM. Bricheteau and Patissier, who had been present at those experiments, had the goodness to comply with the request of the committee, and, conjointly with the reporter, certified by a document which they signed, that it was the same person who had been the subject of the experiments made in the Hôtel-Dieu in 1820, and that they perceived no change in her beyond that which indicated a sensible amelioration of her health. The identity having been thus verified, Mademoiselle Samson was magnetised by M. Dupotet, in presence of the committee. The manipulations had scarcely commenced, when Mademoiselle Samson became restless, rubbed her eyes, expressed impatience, complained, coughed with a hoarse voice, which recalled to the recollection of MM. Bricheteau, Patissier, and the reporter, the same sound of voice which had struck them in 1820, and which then, as upon the present occasion, pointed out to them the commencement of the magnetic action. Soon after, she stamped with her foot, supported her head upon her right hand, which rested upon the elbow-chair, and appeared to fall asleep. We raised her eyelids, and perceived, as in 1820, the ball of the eye turned convulsively upwards. Several questions were addressed to her, and remained unanswered; then, when new ones were put, she exhibited signs of impatience, and said with ill humour that we ought not to annoy her. At length, without having intimated his intention to any one, the reporter threw down upon the floor a table and a billet of wood which he had placed upon it. Some of those present uttered a cry of terror,—Mademoiselle Samson, alone, heard nothing, made no sort of motion, and continued to sleep after as before the sudden and violent noise. She was awakened, four minutes afterwards, by rubbing her eyes in a circular manner with the thumbs. The same billet was then suddenly thrown upon the floor; the noise made her start now that she was awake, and she complained bitterly of the fright that had been given her, although six minutes before she had been insensible to a much greater noise."

The evidence I shall next adduce is that of M. Bouillet, the Professor of Philosophy at the College of Sainte-Barbe, whose knowledge of the metaphysical, as well as the physical sciences, is well known, and whose moral integrity is above all suspicion. In his letter, dated September, 1823, which was published and widely circulated, he gives the following details:—

"Several persons having expressed a desire to witness some magnetic phenomena, after having collected about twenty witnesses, I brought the somnambulist into their presence. This *séance* was nearly a repetition of the same boisterous scenes which had before occurred at the Hôtel-Dieu; and every possible means was had recourse to for the purpose of making the patient hear others, and preventing her hearing me. She was tormented in a thousand ways without effect, when a young man who was present, having provided himself, unknown to me, with a pocket pistol, with the view of making a decisive experiment, suddenly and unexpectedly fired it off close to her ear. Every person present started, and several ladies, taken by surprise, screamed out violently; but the somnambulist was not interrupted in quietly continuing a sentence which at the moment she was addressing to me. It should be added that the pistol was fired off so close to her ear that the bonnet and cap of the poor girl were scorched, and some of the powder lodged under the contused cuticle, yet did she remain perfectly insensible, although, on being awakened, the sensibility which had been temporarily suspended returning, she felt the most acute pain in the neck, and then discovered, with indignation, the state into which, to my deep regret, she had been thrown, and from which, for upwards of a fortnight, she suffered severely."

To this testimony of M. Bouillet, may be added that of M. Husson, President of the Academy of Medicine, who, in his address to that learned Society, observes,—"The operator has succeeded, during this singular state (somnambulism), in producing paralysis, or so entirely closing the senses against all external impressions, that a bottle containing several ounces of concentrated ammonia was held immediately under the nostrils, for five, ten, fifteen minutes, or more, without producing the slightest effect; without impeding respiration, or provoking sternutation. The skin was perfectly insensible, even when pinched so as to leave black marks; nay, it was absolutely insensible to the burning of moxa, and to the acute irritation of hot water strongly impregnated with mustard, although when the skin recovered its normal state, the pain thereby caused was intensely felt." The authority of Dr. Bertrand may be cited to the same effect. "I saw," he observes, "insensibility so decidedly manifested in certain magnetic somnambules, that it could be proved by the most conclusive tests. I saw a magnetiser who used to put his somnambules to sleep before a numerous assembly, and entreat every person present to provide himself with a pin, and thrust them all unexpectedly, at the same moment, into different parts of the body. Accordingly, while the somnambulist was on one occasion singing, forty or fifty pins were thus thrust simultaneously into his flesh without causing the least appreciable alteration in his voice."

In the Memoirs of the Académie des Sciences, page 409, we find a dissertation from M. Sauvage-de-la-Croix, on the somnambulism of a girl of Montpellier, wherein the same phenomenon of insensibility is also attested. "I was visiting the ward of the hospital, as usual, at ten o'clock in the morning, and found the patient in bed. She was talking with more vivacity and spirit than she was ever known to possess in her waking state, passing from one subject to another, and seeming to address several of her friends who stood round the bed. What she said seemed to have some connexion with that which she had repeated in her paroxysm the previous day, in which she recited word for word, an instruction, in the form of a catechism, which she had heard the evening before, and in which she made malicious allusions to the moral conduct of some of the inmates, whom she designated by fictitious names, accompanying the whole with the same gestures and motions of her eyes, which were open, which she would have done in her waking state, yet was she all the while in a profound sleep. This fact was well attested—nobody entertained any doubt of it; but considering that I could never venture to affirm it, unless I applied myself to some experimental test, I made repeated trials on all the organs of the senses while she was talking uninterruptedly. First. As this girl had her eyes open, I thought that simulation, if there were any, could not withstand a violent blow in the face; but this experiment, and a repetition of it, did not make her move a single muscle of the face, it did not even interrupt the thread of her discourse. I tried another experiment, which was, to dart my finger right against her eyeballs, and to hold a lighted candle so close against her eyelids as to burn the eyelashes, but she did not so much as wink. Secondly. A person who was concealed, suddenly uttered a loud shriek close upon her ear, and made a noise by knocking a stone against her bed; but this girl, who at any other time would have trembled with fear, shewed not the slightest sign of hearing. Lastly. I put into her eyes and mouth some brandy and spirit of sal ammoniac; I applied to the eye—the cornea itself—first the tip of a feather, then my own finger, but without any result; snuff blown into her nostrils, pins thrust into her flesh, the forcible contortion of her fingers, had no more effect on her than upon a machine;—she never gave the slightest sign of sensation."

The tenth volume of the Bibliothèque de Médecine contains the memoir of a female somnambulist, who was insensible to the lashes of a whip inflicted on her bare shoulders; and once she had her back well besmeared with honey, and in this state was exposed to the stinging of bees, under a scorching sun, yet although severely blistered, she did not manifest any sign of pain until she was awakened, when she suffered acute agony, and complained grievously of the cruel treatment she had experienced.

This state of complete insensibility is also attested by the Report of the French Academy in 1831, from which I subjoin the following extract :—

"M. Foissac magnetised Cazot, who fell asleep in eight minutes. At three different times, a phial full of ammonia was applied to his nose, but he did not awake. M. Fouquier thrust a pin, an inch deep, into the fore-arm; another was introduced obliquely under the sternum, to the depth of two lines; a third, also obliquely, into the epigastrium; a fourth perpendicularly into the sole of his foot. M. Guersent pinched his fore-arm so as to leave an ecchymosis. M. Itard leaned upon his thigh with the whole weight of his body. They endeavoured to excite a tickling sensation by passing a small piece of paper under his nose, on the lips, the eyebrows, the eyelids, the neck, and the sole of the foot; nothing could awaken him."

It is impossible to conceive any fact more clearly established by human evidence than this perfect insensibility during the magnetic sleep; but it remains for me to add that painful surgical operations have been performed on patients in this state, who during the whole time never manifested the slightest sign of consciousness.

In the "Journal de Toulouse," April 5, 1830, the following case is reported :—"We have already witnessed the curious experiments performed at Toulouse, by the Comte de Brivazac; the document which has now reached us, gives an account of a case which occurred in the department of the Gers, at the house of the justice of peace for the canton of Condom, in the presence of several persons, among whom were some of our own acquaintance. Jean ——, aged twenty-three years, farmer to M. de la Bordère, in the before-mentioned canton, was suffering from an abscess from congestion on the fore part of the thigh. His medical attendants declared that it was necessary to lance it freely, but that the operation should be conducted with much skill and prudence, because the crural artery ran through the tumour, which was very much enlarged. The Comte de Brivazac, whose power as a magnetiser is remarkable, proposed to throw the patient into a magnetic sleep, whereby insensibility might be superinduced, in order that the pain of the operation, inevitable in a waking state, might be spared the sufferer. His proposition was accepted. At the end of two minutes the patient was thrown into the magnetic state. His lucidity was not remarkable. He answered the magnetiser that it was in vain for him to try, as he could not perceive his complaint. At that moment, Dr. Larieu began, with great dexterity, the operation. He repeatedly plunged the probe into the opening made by the bistoury, in order to let out the purulent matter, the flow of which was occasionally impeded by albuminous flakes. A dressing was immediately afterwards applied. During the operation, the patient remained motionless as a statue; his magnetic sleep was in nowise disturbed, and on the proposal being mooted that he should be brought out of the magnetic state, M. de Brivazac spontaneously awakened him. M. Roc then drawing near, asked him whether he would submit to the operation, to which he replied, " I suppose I must, since it is necessary." M. Roc then informed him that it was unnecessary to recommence, as the operation had already been performed. The astonishment of the patient was at its height, when he saw the proofs of it; for he had seen nothing, felt nothing, and remembered nothing, excepting the act of M. de Brivazac laying the palm of his hand upon his forehead to induce sleep."

On the 16th of April, 1829, the following interesting case was reported to the surgical section of the French Academy, by M. Jules Cloquet, the eminent surgeon in Paris, who himself performed the operation. The commissioners of the French Academy, finding, from personal investigation, that the case was fully authenticated, considered it their duty to embody it in their Report. " Madame Plantin, a lady sixty-four years of age, residing at No. 151, Rue St.-Denis, consulted M. Cloquet, on the 8th of April, 1829, for an ulcerated cancer in the right breast, which she had for several years been afflicted with, and which was complicated with a considerable enlargement of the corresponding axillary glands. M. Chapelain, her ordinary physician, who had been in the practice of magnetising her for some months, with the view, as he said, of dispersing the swelling of the breast, had not been able to obtain any other result than that of producing a very profound sleep; during which, sensibility appeared to be annihilated, while her ideas still retained all their clearness. He proposed to M. Cloquet to operate upon her during this state, and as the latter considered the operation indispensable, he consented, and it was fixed for the following Sunday, the 12th of April. During the two days preceding that of the operation, the lady was magnetised several times by M. Chapelain, who, whilst she was in a state of somnambulism, prepared her to submit without fear to the operation, and even brought her to converse upon it with confidence, although, when awake, she rejected the idea with horror.

" On the day appointed for the operation, M. Cloquet, on his arrival at half-past ten in the morning, found the patient dressed, and seated in an arm-chair, in the attitude of a person in a tranquil natural sleep. She had returned, nearly an hour previously, from mass, which she was accustomed to attend at that time. M. Chapelain had thrown her into the magnetic sleep after her return and she then spoke with much composure of the operation she was about to undergo. All the arrangements being made, she undressed herself, and seated herself in a chair. M. Chapelain supported the right arm: the left was allowed to hang down.

" M. Pailloux, *élève interne* of the Hospital St. Louis, was employed to present the instruments, and to tie the vessels. The first incision, commencing at the axilla (armpit), was carried round the upper part of the tumour as far as the inner border of the breast; the second, beginning at the same point, was carried round the lower part of the tumour till it met the first. The enlarged glands were then dissected with precaution, on account of their vicinity to the axillary artery, and the tumour was extirpated. The operation lasted from ten to twelve minutes, and during the whole time, the patient continued conversing tranquilly with the operator, and did not give the slightest indication of sensibility; no motion of the limbs, or of the features; no change in the respiration, or the voice; no alteration even in the pulse could be perceived: the patient never ceased to be in that state of automatic *abandon* and passiveness in which she had been for some minutes before the operation. It was not requisite even to hold, they only supported her. A ligature was then applied to the lateral thoracic artery, which was opened during the extraction of the glands, and the wound being closed by adhesive plasters and dressed, the patient was put to bed, still in a state of somnambulism, in which she was allowed to remain forty-eight hours. An hour after the operation, a slight hæmorrhage appeared, which, however, proved of no consequence. The first dressings were removed on the following Tuesday (the 14th); the wound was cleansed and again dressed, the patient not testifying any sensibility or pain: the pulse preserved its ordinary character. After this dressing, M. Chapelain awakened her, the somnambulic sleep having lasted from an hour before the operation—that is, during two days."

In this case it may be observed that the somnambulic state was maintained for an extraordinary length of time; and then when the patient was awakened, she was again thrown into a similar state. How long this second sleep lasted we are not informed; but this practice I do not hesitate to deprecate. The magnetic sleep, accompanied by insensibility, is restorative to the system, if continued for about two hours; but when it lasts longer, it becomes the source of great excitement, and destroys, rather than restores, nervous energy. From the facts above detailed, it is evident that surgical science should avail itself of this peculiar state to mitigate, or rather supersede, the necessity of inflicting pain during operations; and thus practically applied, magnetism would be of the most essential utility. What is to be the use of this new science?—what good is it to effect? are questions daily asked; but we may rest assured, that however ignorant we may be, at the moment of a discovery, of its ultimate utility, yet every new truth will not fail eventually to admit of some practical application, which will contribute to the benefit, and the happiness, of mankind. It may have been exclaimed against Franklin—What good purpose can be achieved by drawing down the lightning from Heaven by a paper kite? Has not one philosopher been already killed by his presumptuous folly? What advantage can possibly be derived from these Promethean exploits? Yet did this discovery of the electric fluid not only enlarge our views of the constitution of the physical universe, but led to the construction of instruments whereby it admitted of a direct practical application in the treatment of disease which all medical men now appreciate. It is the same with animal magnetism; its direct utility in causing—what might indeed be esteemed a blessing,—an absolute suspension of physical suffering, during otherwise painful surgical operations is, by the preceding cases, distinctly manifest; its curative power in alleviating and removing distressing chronic affections which had resisted every other mode of treatment, has been also clearly established; but it is not possible for us, at present, to raise the veil of futurity, and trace out all the advantages which may hereafter be derived from it.

Hitherto, my attention has been principally confined to the phenomena of insensibility; but the magnetic action induces other physical phenomena, which are many of them very remarkable. One of these I shall here notice. It is this:—The magnetiser in throwing out the magnetic influence, often causes a sort of electrical shock to pass through the body of the patient, who will start as it were convulsively, or a spasmodic contraction of certain muscles will be visible to all the by-standers. A young girl, or rather a child, for she is not twelve years of age, at this moment attends my demonstrations, who is so susceptible of the magnetic influence, that she almost instantly falls asleep, and the approximation of my fingers towards her, causes a short and quick convulsive start, which seems to pervade her whole frame. In another case, a young lady subjected to fits, experiences under the magnetic action, convulsive motions of the shoulders and chest, which are, however, unattended with pain. A gentleman also, who is to me a stranger, has recently attended at my rooms to be magnetised for a paralysis of the left side of the face, which was caused by an abscess which he suffered from five years ago. Under the magnetical action, the muscles of the face, especially those of the paralysed side, over which he has no command, are visibly contracted; the angles of his lips are drawn upwards, and his face assumes almost the character of a mask, and this effect is produced when the magnetic passes are made at a distance of ten or twelve feet from him, even though a screen during the operation be interposed between us.* The

* A medical gentleman, who takes considerable interest in animal magnetism, knowing that Mr. Wright, the patient referred to, was an highly-educated and intelligent gentleman, submitted to him a number of questions, which he succinctly answered, concerning the effects produced, the sensations he experienced, &c. With a copy of these he has favoured me, and I here subjoin them, premising only that I was not at the time acquainted with the parties, and that none of the questions or answers in any

same magnetic effect was produced by me in the presence of the French commissioners, who attested it in their report in the following terms :—" It is chiefly on M. Petit, thirty-two years of age, school-master at Athis, that the convulsive movements have been exhibited with the most precision by the approach of the magnetiser's fingers. M. Dupotet, in introducing him to the commissioners on the 18th August, 1836, announced that he was easily thrown into magnetic sleep, during which state M. Dupotet could at will, without expressing his intention, by the mere approximation of his fingers towards those parts which the commissioners would be pleased to name in writing, bring on

way emanated from me. They were, in fact, only communicated to me while the present sheet was passing through the press.

Q. Are you, independent of the paralysis of the face, in sound health? Have you ever been affected with dyspepsia, palpitations of the heart, or any nervous affection?

A. At present I am enjoying a very sound state of health. When my face was paralysed, about five years ago, I had considerable mental anxiety, and great confinement, with a very high pulse, beating often 125 to 130 a minute.

Q. When the Baron Dupotet magnetised you for the first time were you sensible of any immediate effect during the operation?

A. Immediately upon the Baron Dupotet's commencing his manipulations, I felt physically affected by a stream of coldness.

Q. When the Baron Dupotet commences magnetising you, do you feel nervous, or possessed of any vague apprehension that some mysterious or unknown effect is likely to be produced?

A. Having seen the effects produced on other patients, I was fully prepared to experience the same results, to which I attached notions of pleasure rather than of apprehension. The real effect was completely different from what I had expected.

Q. When the Baron Dupotet passes his hands to and fro before your face, does the monotonous movement before you induce any feeling of *ennui* or mental fatigue?

A. No. The first marked physical sensation is an irritation of the same convulsive movements. The patient was in a short time put asleep, after which, in order to remove every suspicion of any previous understanding between him and the operator, the commissioners handed to M. Dupotet a note, written at the moment, wherein they had specified the parts they wished to be convulsed. Possessed of this instruction, M. Dupotet first directed his hand towards the right wrist, which immediately became convulsed; he then stood behind the patient and directed his finger first towards the left thigh, then towards the left elbow, and lastly, towards the head. Each of these parts were almost immediately seized with convulsive movements. M. Dupotet then

the diseased muscles of the face, and, almost at the same time, a convulsive closing of the eyes. Nearly the same results are experienced when I am magnetised upon the feet.

Q. During the operation do you appear under an influence, which, independent of all such manipulations, sensibly affects you?

A. Yes, entirely so.

Q. When the Baron Dupotet communicates this influence what effect has it on your physical sensations? Do you appear to acquire any new element or principle from without?

A. I appear, as I have before said, to receive a coldness, which quickly operates to the expulsion of heat from the interior; being magnetised for some time, and all things being quiet, this heat is accompanied by great perspiration at all the extremities; for instance, the hands and the feet.

Q. During the operation, and while the paralysed muscles of your face are contracting, are you perfectly conscious of every-thing which surrounds you?

A. I am nearly as conscious as a man ordinarily is with his eyes closed. I cannot speak. Once or twice my consciousness has been much confined to myself and the magnetiser, having forgotten —I do not mean forgotten, but rather being abstracted from— all other objects.

Q. Do the objects and persons about you appear just the same directed his left leg against that of the patient, which became immediately so much agitated that he nearly fell off his seat. He then directed his foot towards the right elbow of M. Petit, which became violently agitated; he then stretched his foot towards the left hand and elbow, and violent convulsive movements developed themselves in the upper limbs. One of the commissioners, M. Marc, with the intention of obviating more effectually every possibility of deception, blindfolded the patient, and the preceding experiments were repeated, with a slight difference in the result. MM. Thillay and Marc directed their fingers toward different parts of the body, and provoked some con-

as when you are not under the magnetic influence, or do they appear as if seen through some different or new medium?

A. I cannot answer this question *experimentally.* All I have to remark upon it is, that I can for the most part see, although dimly, the magnetiser's hand; this sight being rather that of *feeling,* than what we ordinarily term sight.

Q. During the operation, while under the magnetic influence, are you sensible of any exaltation of your mental faculties, either in respect to your perceptions or apprehensions of pleasure or pain?

A. I know of no intellectual exaltation. I am quieted, and the longer the influence is continued, the more calm I become.

Q. When the Baron Dupotet magnetizes you at a distance of six, eight, or twelve feet, or through a screen, is the effect which he produces equal in intensity to that which you experience when he is immediately before you?

A. Yes. There seems to be no difference in the effects resulting either from distance or from interposed objects.

Q. When the Baron Dupotet has established the magnetic influence, are you sensible of any loss of self-command? do you feel in a manner subdued and passive under his control?

A. Partially so. As I have said, I cannot speak, nor when very powerfully influenced, have I been able to move. This is the physical influence. Intellectually I do not feel the magnevulsive movements; so that M. Petit always experienced, on the approach of their fingers, some convulsive movements, which were, however, less promptly developed, and more feeble. This occurred whether his eyes were bandaged or not, and these convulsive movements were more marked when the parts operated upon were submitted to the action of a metallic rod, whether in the shape of a key or the branch of a pair of spectacles."

Such are some of the more manifest, or outwardly visible, effects of the magnetic action; others of a more occult nature may also be produced; but these

tiser more powerful than myself. Morally, or feelingly, I am conscious of a sympathetic *peacefulness* being controlled, but not *forcefully*.

Q. Can you describe the kind of power he exerts over you? Does he in the magnetic state appear to you endowed with any preternatural influence, which exalts him visibly in your imagination?

A. I think I can, though I fear hardly intelligibly. An involuntary assent to the propriety of a suggestion, if transferred from the intellectual to the physical or to the moral, will describe it. There is a physical and moral passivity to a physical and moral actor.

Q. You have heard of *fascination*, in the ordinary sense of the word? Does he rivet your attention so that you cannot escape from the influence? or do you intentionally yield yourself to it?

A. The word fascination aptly describes the influence which the magnetiser exerts. With me it is not an intellectual fascination, but only physical and moral.

Q. Have you ever tried by a determined effort of will to resist the operation? and if so, what has been the result?

A. Yes; and the result has been very painful; great exhaustion being produced, and prostration of the energies. The breathing was painfully affected, and a cold clammy sweat ejected from the extremities, very different from the glowing perspiration that I am ordinarily sensible of. I felt considerable agitation and are so evident to every observer, that, when duly authenticated, their existence cannot reasonably be disputed. It is, after all, be it again observed, only a simple appeal to facts, and that, too, a description of fact so distinct and tangible, that their validity may be tested by every conscientious mind. Finally, it is desirable that these physical manifestations should be thoroughly sifted and determined before proceeding further in this investigation, because they are initiatory to phenomena of a much higher order, which certainly cannot be appreciated if the more subordinate steps of the induction be left imperfect.

loss of self-control, without finding any peaceful influence in its stead.

Q. How many times have you submitted to the operation? Do you think the effect of the magnetic influence increases with every successive sitting?

A. Sixteen times, I think. The influence perceptibly increases, affecting me now more *radically*, while in the first instance my face was really pained by external contortions. The muscles are now more moved, but less apparently on the surface.

Q. Has the magnetic treatment at all affected your general health? Does it invigorate the system, or induce lassitude?

A. I am not sensible of any difference in my general health. I have not slept very well for a few nights, but that might perhaps be attributable to other causes. My friends have remarked a degree of nervous irritation which is not altogether natural to me.

Q Do you think the paralytic affection of your face better for the magnetic treatment you have undergone?

A. Apparently there is but little difference. I seem to think there is a little more vitality in it.

Q. Has the operation of animal magnetism ever given rise to any unpleasant or disagreeable sensations? How do you feel after it has terminated?

A. I am generally affected with a most comfortable perspiration. I am pleased to be magnetised rather than otherwise.

CHAPTER III.

THE PSYCHICAL PHENOMENA OF ANIMAL MAGNETISM.

In all ages, the human mind has been a perplexing problem for the study of philosophers, who have narrowly watched its development in health, its aberrations in sickness, and its occasional unclouding in the hour of approaching death.* Still, its laws are only imperfectly appreciated, because the phases through which it passes are constantly obscured by implication with mere physical effects, inasmuch as the immaterial is so intimately blended with the material portion of human existence, that the most discriminating observer can scarcely distinguish between phenomena that are in themselves purely mental, and phenomena which result from organic action. In other words, man is a being partly physical and partly psychical,—that is to say, an admirably-adjusted physical structure is imbued with a fine spiritual power,

* The unclouding of the mind previous to death, or the prevision of the dying, is a phenomenon manifestly identical with the clairvoyance, or lucidity, of the magnetic somnambulist. Thus does the study of animal magnetism, as we go deeper and deeper into its apparent mysteries, assume a peculiarly sacred interest; it is the unveiling to us of our spiritual nature, and leads us onward even to the verge of that future state of existence, which all men, as they approach, even the most shallow Pyrrhonists, contemplate with a feeling of awe, not unmingled with apprehension.

which controls all its movements ; sustains an exquisite harmony between its different parts ; and establishes, through the *media* of the senses, an almost infinite variety of relations with the surrounding world. The mere automatic mechanism of human existence is thus animated by a spirit-breathing intelligence within, which is necessarily embarrassed and restricted in all its manifestations by the obstruction of those physical conditions under which it is constrained to develop itself. Hence the conflict or antagonism so continually observed between mind and matter. Such is man in the ordinary or waking state. But during sleep he is presented to us under another aspect. His athletic limbs are thrown into repose, the organs of his senses are closed against all external impressions, and in proportion as his physical being sinks into deeper and deeper unconsciousness, the psychical power appears to awaken and energise itself within ; dreams, which memory fails to register, now crowd upon his brain —he becomes restless, speaks, moves, and performs a variety of actions, with a precision and even intelligence which he could not in his waking state have commanded.

Such is the natural somnambulist. It may be a mysterious state of being, but it often occurs, nay, all persons are more or less somnambulists ; we all during sleep occasionally shift our positions, carefully adjust the pillow or bed-clothes to make ourselves comfortable, and perform a variety of trifling actions, of which, at the time, we are perfectly unconscious ; yet these are, in reality, only incipient degrees of somnambulism. Again, many of us talk in our sleep, suddenly start up, hold long discourses, and converse familiarly with those who are supposed to be present, and as the somnambulic state becomes more perfect, phenomena of a more remarkable character appear.

Hence sleep, the repose of our physical being, is always a state bordering on somnambulism, and as it gradually developes itself, the organs of the senses become insensible to all external impressions ; the eyelids are generally closed, or if open, which often happens, the pupil is insensible to light ; the ear, too, may admit the undulations of air, yet it remains deaf to every sound excepting to the voice of the persons with whom the somnambulist may be *en rapport*, or in direct mental communication ; but notwithstanding all this, the mind, independent of the instrumentality of its physical organs, takes a clear and direct cognizance of the relations of surrounding objects. Hence, such persons will in profound darkness, rise, dress themselves, sit down to study, and perform a variety of actions requiring much dexterity and skill, with unerring precision.

A somnambulist mentioned by Gassendi used to rise, dress himself in his sleep, go down to the cellar, and draw wine from a cask. He appeared to see as well in the dark as in daylight ; but if he awoke in the cellar, or elsewhere, he had to grope his way back again. One night he carried on his head a table covered with decanters, up a very narrow staircase, threading his way along to the intended spot without coming in contact with surrounding obstacles ; and this he did with more dexterity than he could have done when awake. In the Bibliothèque de Médecine we meet with the case of a natural somnambulist, who got out of bed in the middle of the night, and went into a house in the neighbourhood, which was in ruins, and of which the bare walls, with a few insecure rafters running between them, alone remained. He, however, climbed to the top of the house, and clambered about from one beam to another, without once missing his hold. In the same work, another somnambulist is described, who dressed himself during the night, put on his boots, fastened his spurs, sprung astride the ledge of a window five stories high, and, fancying himself on horseback, exerted all the energies, motions, and gestures, of a postilion. In this state, persons have been observed to expose themselves in the most perilous positions with impunity; for when the somnambulic state is perfect, there is no apprehension of danger, the gravitation of the body is maintained on the most slender basis, and every motion is determined by a precision so exact and unerring that no unhappy consequences may be apprehended; but if the somnambulist be suddenly awakened this self-possession is entirely lost, and the result proves fatal.

Only recently, at Dresden, a young lady was observed walking in a somnambulic state on the roof of a house, and an alarm being given, crowds of people assembled in the street, and beds, mattrasses, &c., were laid in the street in the hope of saving her life, in case of her anticipated fall. The poor girl, unconscious of danger, repeatedly came forward to the very edge of the roof, smiling, and bowing to the multitude below, and occasionally arranging her hair and dress. The spectators watched her movements with intense interest and painful anxiety, and after thus moving along unconcernedly for some time, she proceeded towards the window from which she had made her exit; but as she approached it, a light, which had been placed there by her distressed family, startled, and suddenly awakened her; upon which, losing all self-command, she fell instantly into the street, and was killed on the spot.

The psychical phenomena manifested during natural somnambulism may be all of them curious; but one, which has been already adverted to, and which is perhaps more remarkable than any other, is that of the transference of the functions of the organs of sense; thus the faculty of vision is proved most distinctly to exist independent of the eye itself,—that is to say, while the eyes are closed, nay, through intervening opaque substances the mind of the somnambulist still perceives the most minute relations of material objects. In the thirty-eighth volume of the encyclopedia edited by Diderot d'Alembert, &c., an interesting case, in illustration of this fact, is reported, on the authority of the Archbishop of Bordeaux. It is that of a young ecclesiastic who was in the habit of rising during the night, in a state of somnambulism, and writing his sermons. When he had finished one page of his manuscript, he would read what he had written aloud, and revise it. In so doing he made use of the expression "*Ce divin enfant,*" and in reading over the passage he changed the word *divin* for *adorable*, and then observing that the pronoun *ce* could not correctly stand before *adorable*, he added to it the letter *t*. In order to ascertain whether he made any use of his eyes, the archbishop held a piece of pasteboard under his chin to prevent his seeing the paper before him; but he continued to write on without being at all incommoded. He copied pieces of music while in this state, during which his eyes were observed to be perfectly closed. It also happened that the words were written in too large a character, and did not stand over the corresponding notes; he soon perceived the error, blotted them out, and wrote them over again with great exactness. Another somnambulist was observed by his physician, Francesco Soave, during sleep, in the act of translating from Italian into French; he looked out for the words, of which he required the meaning, in a dictionary, with as much attention as if awake. On one occasion, his candle was blown out by some of those who watched him, upon which, although other lights were in the room, he immediately rose and went into the kitchen to light it again, for he could not see excepting with the candle he had himself lighted. This is a curious fact, shewing that he was insensible to every other light excepting that upon which his attention was engaged.

The sense of hearing in natural somnambulism undergoes a similar abolition; or rather, is equally abnormal. Such persons are insensible to the loudest noises, and yet hear at a considerable distance the voices of those with whom they are in communication. It is related by Soave, that the somnambulist just referred to, heard the conversation, which appeared to be in conformity with the train of his ideas; but he heard nothing of the discourse which even the same persons held on other subjects. In the transactions of the medical society, at Breslau, the case of a somnambulist is reported, who, during the paroxysm, could not see when his eyes were forced open—could not smell the most volatile spirit, nor even hear the report of a pistol fired close to him. The sense of feeling would also appear to undergo a similar modification. It is stated by Muratori, that Negretti, a remarkable somnambulist, when struck a blow with a stick on the leg, fancied a dog had touched him, and scolded the animal. On being again struck, he threw a bit of bread to the supposed hound, and called to him by his name. It is further reported of Signor

Augustin, an Italian nobleman, who was given to somnambulism, that his servants could not arouse him from his paroxysm by tickling the soles of his feet, or blowing a trumpet in his ear.

Hence, during natural somnambulism, it would appear that the ordinary channels of sensation are entirely closed; the somnambulists see not with their eyes, they hear not with their ears; but a peculiar and new mode of perception exists. It is also well known that such persons, during their somnambulism, often manifest a knowledge and an intellectual activity which they do not at other times possess. Operas have been composed, and poems written, during this state,* and it is a matter of common observation that school-boys will often rise in their sleep, proceed to the school-room, and succeed in making translations and solving problems which they could not in their waking state accomplish. Henricus ab Heer relates the case of a student, at one of the universities in Germany, who having been very intent on the composition of some verses which he could not complete to his satisfaction, rose in his sleep, and

* "In one of my dreams, (says Voltaire,) I was supping with M. Touron, who composed both the words and music of a piece of poetry, which he was singing to us. In my dream I made these four lines for him.

"Mon cher Touron, que tu m'enchantes,
Par la douceur de tes accents!
Que tes vers sont doux et coulants!
Tu les fais comme tu les chantes.

"In another dream, I rehearsed the first canto of the *Henriade* quite different from what it is; I dreamed that verses were being sung to us at supper. Somebody said that he had too much wit, I answered that verses are a festival given to the soul, and that ornaments are required at festivals."

opening his desk, sat down to renew the attempt. He succeeded, and after repeating them aloud, returned to his bed. The faculties, says Richeraud, brought into action by dreams, can lead to a certain series of ideas which we never could have reached in our waking state, and hence mathematicians have, during their sleep, worked out the most intricate calculations, and solved the most difficult problems.

An English physician, Dr. Sibley, relates the following case:—"It lately happened," says he, "that a young gentleman, about fifteen years of age, from one of the public schools, slept in the same room with me. He chose to go to bed early, and when I came into the same apartment about two hours after, he appeared remarkably intent upon his studies though fast locked in the arms of sleep. I stood some time at his bed side and heard him repeat several lines from Homer and Virgil, after this, he repeated, with a bold and nervous accent, the whole of the Hebrew alphabet, then turning, seemed to fall into a more composed sleep. The next morning, at breakfast, I related this circumstance to the company in the presence of the young gentleman, and all were instantly commending the great progress he had made in his studies. The young man instantly declared, that however conversant he might be with Virgil and Homer, he had never heard the Hebrew alphabet repeated, nor did he ever know the name of any one of its characters."*

"It is very remarkable," says Stelling, in his Pneumatology, "that somnambulists who have often been in this state at length attain great clearness of vision; arise, perform all sorts of work, play on an instrument, if

* A Key to Physic and the Occult Sciences, by E. Sibley, M.D., F.R.H.S. London.

they have been taught music, go out to walk, &c., without their bodily senses having the smallest perception of the visible world; they are then in the state of common sleep-walkers. Thus it happened that while I was at Bremen, in the autumn of 1798, a young woman came to ask advice of me respecting her eyes; she was a somnambulist, and had herself decided upon consulting me in the crisis; her mother accompanied her, but she awoke in my presence, and I was therefore obliged to prescribe the appropriate medicines without her assistance."

"We are," observes Sir Thomas Brown, "somewhat more than ourselves during our sleep, and the slumber of the body seems but to be the waking of the soul. It is the ligation of sense, but the liberty of reason and our waking conceptions do not match the fancies of our sleep. I am no way facetious, nor disposed for the mirth and galliardize of company, yet in one dream, I can compose a whole comedy, apprehend the jests, and laugh myself awake at the conceits thereof. Were my memory as faithful as my reason is then fruitful, I should never study but in my dreams, and this time also would I choose for my devotions. But our grosser memories have then so little hold of our abstracted understandings, that they forget the story, and can only relate to our awakened souls a confused and broken tale of that which hath passed. Aristotle, who hath written a singular tract on sleep, hath not, methinks, thoroughly defined it; nor yet Galen, though he seems to have corrected it, for those noctambules, or night-walkers, in their sleep do yet enjoy the action of their senses; we must therefore say that there is something in us that is not under the jurisdiction of Morpheus, and these abstracted and ecstatic souls do walk about in their own corpse, as spirits with the bodies they assume, wherein they seem to hear, see, and feel, though, indeed, the organs are destitute of those faculties which should inform them. Thus it is observed, that men sometimes before the hour of their departure do speak and reason above themselves, for then the soul begins to be freed from the ligaments of the body, and to reason like herself, and discourse in a strain above mortality. They term sleep a death, yet it is waking that kills us. It is that death by which we may be said to literally die daily; a death, which Adam died before his mortality; a death, whereby we live a middle and moderating point between life and death; in fine, so like death that I dare not trust it without my prayers."*

When natural somnambulists are awakened from this state, it is a curious fact, that they remember nothing which has taken place during their somnambulism; but when they again fall into the same state, they then recollect everything which occurred in their former fits. The links of the chain are, as it were, re-united, and the memory of one somnambulic state is as continuous with that of the next as if no interval had elapsed between them. In Moritz' Psychological Magazine, Ritter relates the case of a boy ten years of age, who became subject to fits of drowsiness, and fell suddenly asleep whether standing or sitting. In this state he would converse with persons around him, and although his eyes were completely closed, he was able to see and discriminate all objects presented to him. When awakened, he would recollect nothing that had occurred during his sleep, but would talk of

* Religio Medici, 2nd part, sect. ii. 12. London, 1656.

other matters. On his again falling asleep, he would resume the thread of his discourse where it had been previously interrupted; and again, when he awoke, remembered nothing of the conversation which had taken place during his sleep, although he recollected what had been last said to him when awake. The rule indeed, is, that if somnambulists remember, either distinctly or indistinctly, anything which occurred during their somnambulic state, their somnambulism has been imperfect; and so also in ordinary sleep, the non-recollection or the recollection of our dreams may be esteemed a criterion of our sleep having been more or less profound.

It would be easy to adduce many other interesting and well-attested cases of natural somnambulism; but it is superfluous to accumulate further evidence. From the facts which have been already detailed, it will appear that the following are some of the more obvious psychical phenomena which natural somnambulists exhibit :—

1st. They converse clearly and intelligently with all those persons with whom they are *en rapport*, or in mental relation.

2nd. They perceive the relations of external objects through some other channel than the organs of sense, through which such impressions are usually conveyed.

3rd. Their perceptions in regard to the objects of their attention are more than ordinarily acute; but the organs of the senses are closed against other impressions.

4th. They manifest a clearness or lucidity of ideas, and a temporary knowledge and intellectual activity, beyond that which they possess in their ordinary waking state.

5th. They forget when they are awakened every thing which may have taken place during their somnambulism; but on returning into the same state, they recollect everything which occurred during their former fits.

These, be it observed, are the characteristic phenomena of natural somnambulism, deduced from cases which are thoroughly well authenticated: true, they may be startling, but it should be remembered that such are not to be repudiated because they appear extraordinary, nor are they to be disputed because they cannot at once be satisfactorily explained. It is essentially necessary to distinguish between facts and the explanation of facts; the one is purely a matter of evidence, the other is an affair of theory. Yet are persons generally inclined—nay, some conceive it an intellectual virtue, to deny the existence of the facts which they do not comprehend. This is a manifest absurdity. We believe in our own existence, yet how little do we in reality understand about it. We speak learnedly of the organs which compose the structure of the animal economy, yet how little is known of the mode in which any one of them performs its functions. Furthermore, we discourse in a lofty, philosophical tone, about the attraction of the heavenly bodies, and the laws of gravitation; yet if the astronomer, in his pride of science, will only bend his eye from heaven to earth, he will find himself unable to explain the simple fact, that one piece of inert matter shall affect another. He may observe accurately enough the conditions under which the phenomenon may appear, but in what the attraction itself consists, he cannot explain. In truth, almost the whole amount of human knowledge rests on the existence of facts which are not understood:

we may arrange them with admirable precision, in the order of their development, but the intrinsic relation that subsists between them; the connecting link between cause and effect, cannot be demonstrated. Here our researches end. The inquiring mind, in endeavouring to ascertain a something beyond the sphere of its comprehension, becomes embarrassed;—a finite being cannot understand that which is infinite, but must recoil on the brink of that gulf which appears to divide the Creator from the created.

But to return. It was the Marquis de Puységur, who first induced, or rather, observed particularly, that peculiar state of being which, in contradistinction to natural, is termed magnetic somnambulism. He had been a pupil of Mesmer, and having retired to his estate at Busancy, near Soissons, practically adopted the instructions of his preceptor, and treated magnetically the sick persons who applied to him. He was a man universally esteemed for his integrity and philanthropy; he was also a powerful and efficient magnetiser, and the cures which he wrought were so marked, and so numerous, that the peasantry thronged from all parts of the neighbouring country to consult him. On one of these occasions, he discovered the psychical phenomena referred to; and so marvellous did they appear, even to him, that in transmitting his memoir to the secretary of the Harmonic Society, at Strasburg, he could not refrain from observing, "The time is not yet ripe for the publication of the facts I have witnessed; they are difficult to believe, notwithstanding the mass of evidence which attests them; I therefore beg you will not lend this memoir to any one, for I confide my account of these cases to you only, in order that they may assist you in your reflections, and facilitate you in the means of curing, even more successfully than I have done, during your magnetic trials." It appears, that while magnetising his gardener, he observed him fall into a deep and tranquil sleep; and it then occurred to him, that he would address some questions to him, as he might have done to a natural somnambulist. He did so; and the man immediately answered him with much intelligence and clearness, upon which he persevered in the magnetic operation, and soon found that he possessed an extraordinary psychical influence over him; that all further manual movements were unnecessary; and that without speaking he could mentally communicate with, and control, his ideas. "It is from this simple man," says he, "this tall and stout rustic, twenty-three years of age, enfeebled by disease, or rather, by sorrow, and, therefore, the more pre-disposed to be affected by any great natural agent,—it is from this man, I repeat, that I derive instruction and knowledge. He is no longer, when in the magnetic state, a peasant, who can hardly utter a single sentence,—he is a being, to describe whom I cannot find a name. I need not speak, I have only to think before him, when he instantly hears, and answers me. Should anybody come into the room, he sees him, if I desire it, and addresses him, and says what I wish him to say; not, indeed, exactly as I dictate to him, but as truth requires. When he wants to add more than I deem it prudent strangers should hear, I stop the flow of his ideas, and of his conversation, in the middle of a word, and give his thoughts quite a different turn." He then adds: "I know of no subject more profound, more lucid, than this peasant in his crisis. I have several patients approaching his state of lucidity, but none equal him."

His delight and enthusiasm on making this discovery were indescribable ; " J'étois exalté au dernier point," says he, and he adds, that he might have fancied himself almost favoured of Heaven, had he not recollected the ungenerous opposition which all the scientific academies in Europe had urged against Mesmer and those who had embraced his doctrine. In a letter to his brother, dated May 17, 1784, he observes—" If you do not come, my dear friend, you will not see my extraordinary man ; for his health is almost quite restored. He has, however, told me, when in a crisis, that he should still need to be touched, and pointed out to me the days, Thursday, Saturday, and Monday, for the last time, &c. I continue to make use of the happy power which I owe to M. Mesmer ; and every day I bless him, for I am very useful, and produce many salutary effects on all the sick in the neighbourhood. They flock round my tree ; there were more than one hundred and thirty of them this morning. There is a continual procession in the country. I pass two hours at my tree every morning. It is the best *baquet* possible ; not a leaf of it but communicates health ; all feel more or less good effects from it. You will be delighted to see the picture of humanity which this presents. I have only one regret ; it is, that I cannot touch all who come. But my man, or rather my intelligence, sets me at ease. He teaches me what conduct I should adopt. According to him, it is not necessary for me to touch every one ; a look, a gesture, a wish, is sufficient ; and it is one of the most limited peasants of the country that teaches me this. When he is in a crisis, I know nothing more profound, more prudent, and more clearsighted (*clairvoyant*) than he."

The Marquis de Puységur was now anxious to communicate his discovery to all those who were interested in the science; and among others, M. Cloquet, the receiver of finance, attended to witness his experiments. "Attracted, like others," says he, "to this spectacle, I went, prepared to be a calm and impartial observer, determined to be on my guard against the illusions of novelty and wonder, and firmly resolved both to look and to listen well."

He then describes the proceedings employed by M. de Puységur for acting upon the patients; and having related various scenes of magnetisation, adds—"The consummation of this state (the magnetic state) is an appearance of sleep, during which the physical faculties seem to be suspended, but to the advantage of the intellectual faculties. The eyes of the subject are closed; his sense of hearing is abolished; he wakes only at the sight of the master (*du maître.*) No one should touch the patient during a crisis, not even the chair on which he is seated; it would cause him much suffering and convulsions, which the master only can calm."

"These patients, during the crisis, possess an extraordinary (*surnaturel*) power, by which, on touching a patient presented to them, on passing their hand even over the clothes, they feel which is the affected viscus—the suffering part; they point it out, and indicate pretty nearly the suitable remedies."

"I was touched by a woman of about fifty. I had certainly informed nobody of my particular complaint. When she had looked attentively at my head for some time, she told me that I had frequent head-aches and a buzzing noise in my ears, which is perfectly true. A young man, one of the incredulous spectators of this experiment, submitted to it, and he was told that his complaint was pains in the stomach, and obstructions in the abdomen, that this was in consequence of a disease which he had some years previously, which he confessed to be in accordance with truth. Not satisfied with this divination, he went to another sleeping physician, about twenty yards from the first, and here he was told the very same thing. I never saw any person so completely confounded, for he had evidently come for the purpose of scoffing and contradicting, and not with a view of being convinced."

"Another singularity, not less remarkable than all that I have just disclosed, is, that these sleepers, who during four hours have touched patients, have reasoned with them, remember nothing—absolutely nothing, after the magnetiser has thought proper to disenchant them—to restore them to their natural state. The time that has elapsed between their entering into the crisis and that of their coming out of it, is, as it were, obliterated. Not only has the magnetiser the power, as I have already said, of making himself heard by these somnambulists, but I have frequently seen him point his finger from a distance to individuals while in the crisis, and in a state of spasmodic sleep, and make them follow him wherever he chose; or send them from him, either to their own homes, or to different places which he designed, without telling them,—the somnambulists, it should be remembered, having their eyes the whole time completely closed. I have forgotten to mention, that the intelligence of these patients is singularly susceptible. If, at distances by no means inconsiderable, conversation be held offensive to propriety, they hear it, as it were, internally, their minds are disagreeably affected by it, and complain of the circumstance; and this has several times occasioned scenes of confusion to ill-witted jesters (*pour les mauvais plaisants*), who indulged, at M. de Puységur's residence, in inconsiderate and misplaced sarcasms."

The discovery of magnetic somnambulism, instead of advancing, rather retarded than otherwise the progress of the science, inasmuch as the facts above stated appeared so startling and incredible, that many held it superfluous even to entertain any consideration of them; and dismissed them, at once, as idle fictions and knavish impositions. The concurrent testimony, however, of so many individuals, who had no interest whatever in the matter, and whose integrity and powers of judgment could not be impugned, induced many who were incredulous to examine the facts; and hence observations on this phenomenon have, during the last fifty years, been accumulated in all parts of Europe, even by those very individuals who were originally the most determined and perverse sceptics. I now, therefore, proceed to describe the phenomena which magnetic somnambulists exhibit; but it is proper to observe, that no criterion of comparison can be established between different cases, because men differ as much from one another in somnambulism as they do in their habitual waking state. The Count de Redern, a distinguished savant, who devoted much attention to animal magnetism, observes,—

"The body is more erect than in the waking state; there is a marked acceleration of the pulse, and an augmentation of irritability in the nervous system; the touch, taste, and smell, have become more delicate; the sense of hearing is affected only by sounds proceeding from the bodies with which the somnambulist happens to be in direct or indirect relation,—that is to say, which are in communication through the vital fluid, from having been touched by him and his magnetiser. His eyes are closed, and have no longer the power of vision; but he has a kind of sight, which may be called internal,—that of the organisation of his own body, of that of his magnetiser, and of the persons with whom he is placed in relation; he perceives the different parts of them, but in succession only, and according as he directs his attention to them; he distinguishes their structure, form, and colour. He has sometimes the faculty of perceiving external objects by a peculiar kind of sight: they appear to him more luminous, more brilliant, than in the waking state. He experiences a painful reaction of the sufferings of the persons with whom he is in relation; he perceives their diseases, foresees their crises, has a perception of the suitable remedies, and not unfrequently that of the medicinal properties of substances presented to him. His imagination has a tendency to exaltation; he is jealous, full of vanity and self-love, and disposed to make use of little artifices to give himself importance. His will is not inactive, but it is easily influenced by the magnetiser. Very striking contradictions are observable between his ordinary opinions and those in his state of somnambulism; he condemns his own actions, and sometimes speaks of himself as he would of a third person quite a stranger to him. He expresses himself better, has more intelligence, greater powers of combinations, possesses more reason, more morality, than in his waking condition, all the ideas of which are present to him. When the somnambulist returns to the waking state, he has entirely forgotten all that he had said, done, and heard, during the fit of somnambulism," &c.

" When magnetism produces somnambulism (says M. Husson,) the individual in that state acquires a prodigious extension of the power of sensation. Several of his exterior organs are, so to speak, put to sleep, especially those of sight and hearing, and all the operations dependent on them are performed internally."

" The somnambulist has his eyes closed; he neither sees with his eyes, nor hears with his ears: yet he sees and hears better than a waking person. He sees and hears only those with whom he is in relation. He sees only that at which he looks; and he usually looks at those objects only to which his attention is directed. He is submissive to the will of his magnetiser in all things which cannot injure himself, and in all that does not oppose his own ideas of justice and truth. He feels the will of his magnetiser. He sees, or rather he has a perception of, the interior of his own body, and of that of others; but he usually remarks those parts only which are not in the natural state, and which disturb the harmony of it. He recals to his memory things which he had forgotten in his waking state. He has previsions and presentiments, which may be erroneous in several circumstances, and which are limited in their extent. He expresses himself with surprising facility. He is not free from vanity; his self-improvement is progressive (*il se perfectionne de lui-même*) for a certain time, if guided with discretion; but if ill directed, he goes astray. When returned to his natural condition, he entirely loses the recollection of all the sensations and ideas he had during his state of somnambulism; so that these two states are as entirely strangers to one another as if the somnambulist and the waking man were two different persons."

M. Deleuze, whose important Critical History of Animal Magnetism has been already noticed, gives the following summary of the phenomena of magnetic somnambulism:—

"When magnetism produces somnambulism, the being who is in this condition acquires a prodigious extension in the faculty of sensation; several of his external organs, generally those of sight and hearing, are inactive, and all the sensations which depend upon them take place internally. Of this state there is an infinite number of shades and varieties; but in order to form a right judgment of it, we must examine it in its greatest difference from the state of waking, passing over in silence all that has not been confirmed by experience.

"The somnambulist has his eyes shut, and does not see with his eyes; he does not hear with his ears; but he sees and hears better than one who is awake.

"He sees and hears only those with whom he is in communication, (*en rapport*). He sees nothing but what he looks at, (*ce qu'il regarde*), and he generally looks only at the objects to which his attention is directed.

"He is under the will of his magnetiser in regard to everything that cannot hurt him, and that he does not feel contrary to his ideas of justice and truth.

"He feels the will of his magnetiser.

"He perceives the magnetic fluid.

"He sees, or rather he feels, the interior of his

body, and that of others; but he commonly observes only those parts of it which are not in their natural state, and disturb the harmony of the whole.

"He recovers the recollection of things he had forgot when awake.

"He has prophetic visions and sensations, which may be erroneous in some circumstances, and which are limited in their extent.

"He expresses himself with astonishing facility.

"He is not free from vanity.

"He becomes more perfect of his own accord for a certain time, if guided wisely. He wanders when he is ill directed.

"When he returns to the natural state, he entirely loses the recollection of all the sensations, and all the ideas which he has had in the state of somnambulism, so that these two conditions are as foreign to one another as if the somnambulist and the waking man were two different beings.

"Generally speaking," adds M. Deleuze, "the magnetic somnambule perceives innumerable relations in all objects; he perceives them with an extreme rapidity, and in one minute runs through a train of ideas which to us would require many hours. Time seems to vanish before him; he himself wonders at the variety and rapidity of his perceptions; he is inclined to ascribe them to the inspiration of another intelligence. He now sees this new being within himself; he considers himself, while in somnambulism, as a different being from himself when awake; he speaks of himself in the third person, as of somebody whom he knows, submits to his remarks,—to whom he gives advice, and for whom he feels more or less sympathy. He now hears an intelligence, a soul speaking to him, and revealing what he wishes to know."*

In the Cours de Matière Médicale, by Desbois de Rochefort, published by M. Lullier Winslow, we find a chapter containing sufficient evidence to enable us to form an opinion on magnetism. The following are some few of the author's observations :—

* Virgil represents the Sybil speaking to Æneas, in the same state in which we now see certain somnambules.

" 'Deus, ecce Deus!' cui talia fanti
Ante fores, subito non vultus, non color unus,
Non comptæ mansere comæ, sed pectus anhelum
Et rabie ferâ corda tument, majorque videri,
Nec mortale sonans afflata est numine quando
Jam propiore Dei," &c.—Æneidos, lib. vi., v. 47.

" 'He comes! behold the god!' Thus while she said,
(And shivering at the entry staid,)
Her colour changed; her face was not the same;
And hollow groans from her deep spirit came.
Her hair stood up; convulsing rage possessed
Her trembling limbs, and heaved her lab'ring breast.
Greater than human kind she seemed to look;
And with an accent more than mortal, spoke.
Her staring eyes with sparkling fury roll;
When all the god came rushing on her soul."—Dryden.

The spirit, it is said, moved with so much force those who slept in ancient temples, that they were ravished out of themselves; so that they had divine visions of the most extraordinary nature, and could understand various mysteries. When the spirit began to move them, their countenances gradually altered; their limbs were convulsed; they dropped down as if they had been seized with some falling fit. When lying on the ground, they remained in death-like stillness: sometimes their whole bodies shook in a dreadful manner, at other times they lay still, as if their bodies had lost all power of motion. On awaking from this ecstatic sleep, they related the astounding visions which the spirit had unfolded to them.

" When the Laplanders wish to know what occurs in places

"It is proved," says he, "that magnetism is a principle entirely unknown in its essence, but demonstrated by its astonishing effects; that this impalpable, imponderable principle, the nature of which is unknown to us, though it appears to have some affinity with electricity, is so extremely subtile, that it seems to be transmitted from one individual to another by the mere act of volition; and that, when it

remote from their habitations, they send out their familiar spirit in search of intelligence; and when they have sufficiently excited their own imaginations by the sound of drums and other musical instruments, they feel a kind of intoxication, during which certain things are revealed to them, which they never could have known in their natural state."

Many similar instances may be found in profane history. Among others, the following, taken from a Treatise on the Rites, Ceremonies, and Mysteries, in use among the Brahmins, written long before the expedition of Alexander in Hindostan. Therein it is recorded that, by a common custom of the country, called *matricha machow*, they acquire a new life. They consider the epigastric region as the habitual seat of the soul. They move their hands from this part of the body up to the head; they squeeze and rub certain nerves which, they imagine, correspond with those parts, and pretend that, by such manipulations, the soul is transferred to the brain. As soon as the Brahmin thinks he has reached this state, he believes that his body and soul are reunited to the Deity, and that he has become part of the divine essence.

"Whenever I wish it," says Cardan, "I come out of my body so as to feel no sensation whatever, as if I were in ecstacy. When I enter this state, or, more properly speaking, when I plunge myself into ecstacy, I feel my soul issuing out of my heart, and, as it were, quitting it as well as the rest of my body through a small aperture formed at first in the head, and particularly in the cerebellum. This aperture, which runs down the spinal column, can only be kept open by very great efforts. In this situation, I feel nothing but the bare consciousness of existing out of my own body, from which I am distinctly separated. But I cannot remain in this state more than a very few moments."

does operate, it gives rise to a great variety of phenomena, the chief of which are,—release from coughing, yawnings, a kind of stupor or giddiness, sleep more or less profound, a semi-cataleptic state, convulsions, and, finally, a real state of somnambulism, often attended by a kind of transfer of the senses towards the epigastrium, and by an incredible extension of sensibility.

"It is this somnambulism, an extreme result of animal magnetism, which allows of, and promises success, through its known efficacy in the diagnosis and treatment of diseases. It is by means of this incomprehensible, undefinable development of the general sensibility, that magnetic somnambules acquire, without the help of the senses, not only a knowledge of the surrounding objects on which their attention is directed, either spontaneously or by the will of another, but also the faculty of distinguishing objects removed at a distance, or placed, with regard to the somnambule, beyond opaque bodies; and, by consequence, to know the economy and functions of their own system, or the systems of the individuals presented to them. The most positive, the most authentic, and incontrovertible facts, justify, establish, and guarantee all these phenomena of magnetic somnambulism, and prove that, under certain circumstances, the lucidity of somnambules may be of great succour to determine the seat and the nature of diseases, particularly those which come under the denomination of organic diseases." He then gives the nomenclature of those affections in which it has been applied with success.

Here, also, I may cite a passage from the speech of

M. Chardel, a member of the Académie de Médecine, who, in addressing the Académie during the discussion on magnetism, made the following remarks :—

"Among the phenomena most frequently elicited by the magnetic agency, we reckon,—1. A profound and protracted sleep, constantly preceding and following the production of somnambulism.—2. The exaltation of the intellectual faculties.—3. A perfection of the sight, which enables the somnambule to perceive the magnetic fluid.—4. The faculty of acquiring information on the state of the internal organs," &c. &c.

In a memoir addressed to the Académie de Médecine, with the view of inducing that body to take an interest in magnetism, a physician observes,—"By laying the hand successively on the chest and abdomen of a stranger, my somnambulists immediately discover his or her malady; they moreover indicate whether the cure be possible, easy, or very remote, and what means are to be used, in order to obtain this result by the safest and readiest method, in doing which they never deviate from the established rules of sound medical doctrine. Notwithstanding the apparent magnitude of the promise, I am confident that there is no disease, whether of an acute or chronic character, of a simple or compound nature, not one of those the seat of which lies in one of the three splanchnic cavities, which cannot be detected, and properly treated by lucid somnambules. I have already made a successful application of animal magnetism in the treatment of many diseases which had hitherto been entirely misunderstood, or considered as incurable. I applied it with equal success in ordinary diseases, usually known by their symptoms, progress, and termination; and I invariably found that the information supplied by the somnambules bore the stamp of great sagacity, precision, and unerring judgment."

The most extraordinary phenomenon of magnetic, as well as of natural somnambulism, is clairvoyance, or vision without the use of the eyes; yet no fact has ever been more clearly and indisputably established. It has been already seen, nay, it is a notorious fact, that natural somnambulists enjoy this faculty. It is inexplicable in them—it is equally so in the magnetic somnambulist; but, as I have before stated, we must abide by the evidence of the fact being conclusive, and not wander out of the path of positive observation to pursue any *ignis fatuus* theory which imagination may suggest to explain the mystery. It may, indeed, appear tiresome, my adducing so many authorities, and citing so many facts of the same description; but I have no alternative; I am anxious to show clearly that the weight of evidence in proof of their existence is irresistible; and when I reiterate that I am myself prepared to reproduce the same phenomena, the conditions of the operation being, *ceteris paribus*, similar, surely the most perverse of my antagonists will allow that no fairer appeal can be made to the common sense and judgment of mankind. The cases of vision during magnetic somnambulism without the instrumentality of the eyes are very numerous, and those which I shall here adduce are selected principally on account of their being attested by living authorities—men who hold a distinguished rank in the faculty of medicine, at Paris, and in other enlightened cities.

The first fact I shall adduce is taken from the Gazette de Santé, and recorded in the Journal de Paris of the 24th Brumaire, an. XIV.

"The public journals," says the editor, "are now re-echoing the prodigies of a female somnambule of Lyons, who with her eyes shut can read a sealed letter, tell the thoughts, and give an account of the sensations, &c., of another person. The most singular circumstance is, that this woman is of genteel extraction, and that on account of her superior education and independent fortune, she is far above the suspicion of simulating those extraordinary scenes; in addition to which, the most eminent personages, the most learned physicians and scientific men of Lyons, appear perfectly convinced of the reality of these prodigies."

M. Deleuze, in his memoir on the Lucidity of Somnambules, gives the following relation:—

"A young patient had read to me, with great fluency, seven or eight lines, notwithstanding her eyes were so covered as to be of no use to her, after which she was obliged to stop, saying she felt much fatigued. Some days after, wishing to convince a few sceptics, whom he could not take with him to the somnambule's residence, he presented to her a box of pasteboard, perfectly shut, and in which were written these words—*amitié santé, bonheur*. She held the box in her hand for a long time, experienced much fatigue, and said that the first word was *amitié;* but that she could not read the others. On being urged to make fresh attempts, she consented, and said, as she returned the box, I cannot see it plainly; however, I think that the next two words are *bonté, douceur*. She made a mistake in these two words; but, as we see, these words had a very great resemblance with those that were written, so striking a coincidence could not be attributed to mere chance;" the three written words being, *amitié, santé, bonheur*,—and those mentioned by the patient—*amitié, bonté, douceur*."

The next are more recent, and perhaps, therefore, more conclusive attestations.

M. Rostan, professor of the faculty of medicine, well known by his various writings, which have satisfied public opinion as to his merit, might have apprehended a partial loss of the esteem he enjoyed, by acknowledging the phenomena of somnambulic lucidity. He, however, did not hesitate to do so; and the article *Magnétisme*, in the Dictionnaire des Sciences Médicales, contains several extremely curious facts, similar to those I have now adduced. He quotes the observations made upon a somnambule, who told him accurately and repeatedly, the exact hour, by a watch held at the back of her head. The following is an account of this experiment. After having treated of the somnambulic faculties in general, M. Rostan thus proceeds:—

"But if sight be obliterated in its natural organ, it is satisfactorily demonstrated by me that it may exist in various parts of the body. The following is an experiment I have often repeated. It was performed in the presence of M. Ferrus. I took my watch and held it at the back of the head, at a distance of three or four inches from the occiput. I asked the somnambule whether she saw anything. 'Certainly,' said she, 'I see something shining; it gives me pain.' Her countenance was expressive of pain; and ours bespoke our astonishment. We stared at each other, and M. Ferrus at last broke silence, by observing to me, that if she could see something shine she could probably tell what it was.

"What do you see shining? 'Oh, I do not know I cannot tell.' Look well. 'Why, it fatigues me so... Why, it is a watch.' Fresh surprise on our part. But if she can see that it is a watch, again said M. Ferrus, she will probably tell us the time. Can you tell what time it is? 'Oh, no, that is too difficult.' Pay attention, and look well. 'Why, I will try...I can, perhaps, tell the hour, but I can never see the minutes.' When she had looked with the utmost attention, she said, 'It is ten minutes to eight,' which was then the exact time. M. Ferrus wished to repeat the experiment himself, which he did, with similar success. He altered several times the direction of the hands on his watch; and when it was presented to her, without our having looked at it, she was right every time."*

M. Georget, another colleague of M. Rostan, affirms that he observed a somnambulist who exhibited very astonishing phenomena of prevision and lucidity; "so much so," says he, "that in no work on magnetism, not even in Petetin's, nor among all the phenomena that occurred to my observation, have I ever met with anything more extraordinary."

"All the facts," says Dr. Rostan, "that M. Georget published, I have seen with my own eyes; many of those experiments were performed at my own house; we had both no other end in view beyond that of obtaining information. We were both inclined to doubt."

M. Chardel, also, whom I have already quoted, has recorded, in a work on magnetism published by him four years ago, several instances of the function of sight without the assistance of the eyes.

* Dictionnaire de Médecine. Paris, 1827, Article, *Magnétisme.*

The following is one of the facts he adduces :—

" The somnambule having recovered her senses (for she had just been seized by syncope), called for some water ; I went to take a decanter from the mantle-piece, but it was empty. I took it, for the purpose of filling it, into the dining-room, where I had observed a filtering tank ; I turned the cock, but no water came ; and yet the tank was full. I thought that the cock should first be unstopped, and I did it with a piece of wood which I split off ; but still the water did not come out. I then supposed that the air-hole of the reservoir was obstructed, and as it was very narrow, it was necessary again to split the piece of wood in order to introduce it : but I was not more successful than before. At last I resolved upon filling my decanter with unfiltered water. My somnambule was still in the same attitude in which I left her. She had seen me all the time, had followed all my movements, and detailed them to me without omitting a single circumstance, notwithstanding there was between her and me two walls and a parlour,—and my actions included a number of minute details which nobody could have imagined."* M. Chardel proceeds :—" I might adduce many more instances of similar sight, and even at considerably greater distances ; but the circumstances would not be more conclusive."

Here also I shall subjoin additional attestations by men no less worthy of credit.—M. Francœur, a distinguished mathematician, read in 1826 to the Société Phi-

* " Illi viva acies, nec popula parva, sed ignis
Trajector nebulæ, vasti et penetrator operti est."
Aurelius Prudentius.

No, the sight of the soul is not confined to a narrow retina ; it is a living fire, a lightning which rends the clouds, and darts into the unfathomable depths of unknown worlds.

lomatique a memoir, wherein he stated that "He had been at Aix in company with respectable physicians,—namely, with Dr. Despine, chief physician of the watering establishment, who informed him that he (Dr. Despine) had witnessed, for months together, the singular phenomenon of the transference of the senses, and he thought he would serve the cause of truth by making it known to the Society."

In the first observation of this memoir, we read that the patient had the power of seeing, hearing, and smelling, with her fingers. The second is much more curious; it was made on the daughter of a certain gentleman, who enjoys the esteem of the whole town of Grenoble, where he lives in retirement. He did all in his power to conceal the malady of his daughter, which gave him great pain, and constantly refused to admit visitors inquiring after her.

Among the various states presented by this malady and minutely described by Dr. Despine, he insists, in particular, on somnambulism; and I shall here transcribe, *verbatim*, the most positive passage on the transposition of the function of sight.

"Not only did our patient hear with the palm of her hands, but we saw her read, without the help of her eyes, and with the extremities of her fingers, which she rapidly agitated over the page which she intended to read, and without touching it, as if to multiply the feeling surfaces; she thus read a whole page of a novel by Montholieu.

"At other times we saw her single out of a parcel containing upwards of thirty letters, one which had been pointed out to her, and read on the dial, and through the glass of a watch the hour indicated by the hands; she also wrote several letters, corrected them by a second reading, marking the mistakes as she went on, and recopied one of them, word for word. During all these operations, a thick pasteboard screen intercepted in the most effectual manner every visual ray that might have reached the eye.

"The same phenomenon took place at the sole of the feet, and at the epigastrium; and the patient seemed to experience a painful sensation when simply touched."

Other minutely detailed accounts of sight without the assistance of the eyes, will be found in a memoir by Dr. Delpit, on two nervous affections.* "One of the patients read," says the author, "and that very distinctly, when her eyes were hermetically closed, and by running her fingers over the letters. I made her read printed characters in this manner, both in the open day-light or in the most profound darkness, on opening the first book that came into my hand; and oftentimes written characters, by giving her sundry notes, which I had prepared previous to my coming. Whether the sense of sight was in her case supplied by that of touch, I cannot tell; but I affirm that she read fluently by running her fingers over the letters."

Here is assuredly conclusive evidence, given by a man under whose observation the facts unexpectedly fell, without his seeking them, or even suspecting the possibility of their existence, and who, being struck by their inconceivable singularity, took up his pen to record what he had seen.

In the very short remarks affixed to the preceding observations of M. Delpit, he refers the reader to the authority of many well known writers, and particularly to M. Dumas. "Five years ago," says this celebrated

* Bibliothèque Médicale, t. 56, p. 308.

author, "a young female, who came from the department of Ardèche to Montpellier to consult the physician about an hysterical affection, accompanied by catalepsy, exhibited a most extraordinary phenomenon. She experienced, during the whole period of her attacks, so great a concentration of sensibility towards the precordial region, that her organs of sense seemed as if they had been transported to that part. She referred to the stomach all the sensations of sight, hearing, and smelling, which were then no longer produced through the usual organs." This rare phenomenon, observed in a subject highly interesting in every respect, was a matter of serious attention for the medical profession, and of intense curiosity for the public at large.

"I am fully aware," says M. Delpit, "that facts of this nature, in direct opposition to all the known laws of physical science, could not obtain, without much difficulty and restriction, the sanction of prudent men, who apprehend being led into error. But if we multiply observations on this subject, if we ascertain with scrupulous exactness the least circumstances of each observation, we must finally acknowledge, not only the possibility, but the existence, of a phenomenon which appears so marvellous merely from our not having seen many facts with which it could bear comparison."

These are recent facts, and of such a nature as to preclude the possibility of our being deceived; they are testimonials furnished by living writers, men far above every suspicion of imposture.

It would be difficult, indeed, to discover any motives which could lead to such assertions, had not the facts stated been quite evident. Besides, they are of daily occurrence. The following is an observation published in the Gazette de Santé, of September 1829:

" There is now, in M. Fouquier's wards, a patient in whose case animal magnetism has developed the most curious and, in some points, incredible phenomena. Without siding with the lovers of the marvellous, it is incumbent upon us to record individual facts when duly authenticated. Now, the facts which we are about to relate having been witnessed by a large number of spectators, we presume we may present them to our readers as being entitled to their confidence.

" Petronille Leclerc, twenty-six years of age, seamstress, entered the Charité to be treated for a spasmodic cerebral affection bordering on epilepsy. She was of an exceedingly nervous temperament, very pale, exhausted by former sufferings, and extremely irritable. She had been seized, after a violent fit of anger, with a turning upwards of the eyeballs, in which position they remained fixed, and resisted all sorts of remedies. The idea occurred to M. Sbire, her medical attendant, of trying the application of magnetism. He made the first trial on the 29th of August last, and subsequently repeated it many times. The most remarkable circumstances of this treatment were noted down, and some of them occurred in our own presence. The following is an account of them:—

" At the first sitting, the somnambule gave several proofs of lucidity. The gentleman who had magnetised her presented to her various objects, such as a flask, some sugar, and bread, which she perfectly discriminated without seeing them, for her eyes were bandaged; and, moreover, in order to answer the questions put to her, she generally turned her head on the opposite side, and buried her face in her pillow. Once, without being questioned, she said to the same person, who was holding her hand, 'You have a head-ache,' which was really the fact; but, with the view of perplexing her, the student told her she had made a mistake. 'That is very singular,' said she, 'then I must have touched somebody who has a head-ache, for I felt it well.' She also distinguished several individuals present by certain peculiarities in their dress. One of the most remarkable circumstances is this:—The magnetiser, on retiring, had promised that he would return at about half-past five to wake her. He came before the appointed time, and the somnambule observed to him, that it was not yet half-past five; but he replied, that a letter which he had just received obliged him to return to her. 'O yes,' said she immediately, 'it is the letter now in your pocket-book, between a blue and yellow card;' which fact proved to be perfectly correct.

"A watch was placed at the back of her head, and on being requested to tell the time, she said, 'Six minutes past four;' it was then seven minutes past four."

The 17th number of the same journal contains a note, stating that the same patient had, according to her own prediction, a rather intense inflammation of the mouth and fauces, with ptyalism. It was accompanied by constipation, as she had previously foretold.

These assertions should have some weight, when it is added, that the editors of this paper had long been opposed to magnetism. In their eyes, magnetisers were little better than poisoners.

Here, also, is another extract from the Gazette Médicale de Paris, dated October 2, 1832 :—" There is now observed, at the Hospital Della Vitta, in Bologna, a very extraordinary phenomenon of animal magnetism. In this hospital is a patient, who is seized every third day, precisely at eleven in the morning, with a convulsion so violent that he loses entirely the faculty of perceiving sensations. Sight, hearing, and smelling, totally disappear; the organs of the senses cease their functions; both his hands are so firmly clenched that it is impossible to open them, and by using more force the fingers would certainly break. But Dr. Cini, son of the artist of that name, under whose care he was placed, discovered, after long and attentive observations, that the epigastrium,* at about one inch above the umbilicus, received, during the convulsive fit, all the perceptions of the senses, so as to replace them to a certain extent. If we speak to the patient, touching the epigastrium with the finger, he answers immediately; and if requested to do so, he will of his own accord open his hands; if a body be placed upon the epigastrium the patient describes its shape, smell, quality, and colour. During the contact of the finger the convulsion gradually diminishes, and seems to die away; but if the finger be placed upon the heart, the convulsion revives, and lasts as long as the finger remains in this position. If a flute be played upon while touching the epigastrium, the patient hears the music, and when, without interrupting the performance on the instrument, the finger leaves the epigastrium to touch the precordial region, and then returns to the epigastrium, the patient in-

* By *epigastrium* the unprofessional reader should understand the pit of the stomach.

quires why the musical performance is now and then interrupted. These experiments were performed in the beginning of September, in the presence of both professors and students; and they excited the greatest surprise."

In a curious work, printed in Germany some years ago, another interesting case of the same description is related. The author was an eye-witness himself of the facts, and in addition he brings forward the experiments of three eminent physicians, whose accounts of the effects they produced agree in every respect with his own. In the case he treated, the somnambulist saw a paper written by the Baron de Strombeck; and stated that the writing consisted of two paragraphs, containing a certain number of lines. "I went for the paper," says he, "and counted the lines, and I shuddered as if I had seen a supernatural being." The same ecstatic patient saw in a room, in an upper story, a proof sheet from the printer laid on M. Strombeck's bureau, when he himself was not aware of its having been brought. She pointed out the particular situation of objects placed behind her, although they had been purposely deranged from their usual position. She told to a minute the hour indicated by Dr. Schmidt's watch, which differed from every other in the room. She said that Dr. Marcard had some money about him in a long green purse, such as were then quite out of fashion, upon which it appeared that he had been wearing this purse a very few days, and he affirmed that he had never taken it out of his pocket, excepting when at home.

The clairvoyance of magnetic somnambulists was also unequivocally attested by the conjoint evidence of the commissioners of the French Academy. They carefully examined the fact, and in their report give the following details. I was, it will be observed, in the first case, myself the operator. On the 18th of March, 1826, at half-past eight in the evening, M. Petit was set asleep in one minute.

"The president of the committee, M. Bourdois, ascertained that the number of pulsations diminished at the rate of twenty-two in a minute, and that there was even some irregularity in the pulse. M. Dupotet after having put a bandage upon the eyes of the somnambulist, repeatedly directed towards him the points of his fingers, at the distance of about two feet. Immediately a violent contraction was perceived in the hands and arms towards which the action was directed. M. Dupotet having, in a similar manner, approximated his feet to those of M. Petit, always without contact, the latter quickly withdrew his. He complained of great pain and a burning heat in the limbs, towards which the action had been directed. M. Bourdois endeavoured to produce the same effects; and he succeeded, but less promptly, and in a more feeble degree. This point being established, we proceeded to ascertain the lucidity (*clairvoyance*) of the somnambulist. He having declared that he could not see with the bandage, it was taken off; but then we determined to assure ourselves that the eyelids were exactly closed. For this purpose, a candle was almost constantly held, during the experiments, before the eyes of M. Petit, at a distance of one or two inches; and several persons had their eyes continually fixed upon his. None of us could perceive the slightest separation of the eyelids. Mr. Ribes, indeed, remarked that their edges were super-imposed, so that the eye-lashes crossed each other.

" We also examined the state of the eyes, which were forcibly opened without awakening the somnambulist; and we remarked that the pupil was turned downwards, and directed towards the great angle of the eye. After these preliminary observations, we proceeded to verify the phenomena of vision with the eyes closed.

" Mr. Ribes, member of the Academy, presented a catalogue, which he took from his pocket. The somnambulist, after some efforts which seemed to fatigue him, read very distinctly the words, *Lavater. Il est bien difficile de connaître les hommes.* The last words, were printed in very small characters. A passport was placed under his eyes; he recognised it, and called it a *passe-homme.* Some moments afterwards, a *port-d'armes* was substituted, which we all know to be in almost all respects similar to a passport, and the blank side of it was presented to him. M. Petit, at first, could only recognise that it was of a particular figure, and very like the former. A few moments afterwards, he told us what it was, and read distinctly the words, *De par le roi*, and on the left, *Port-d'armes*. Again, he was shewn an open letter; he declared that he could not read it, as he did not understand English. In fact, it was an English letter.

" M. Bourdois took from his pocket a snuff-box, upon which there was a cameo set in gold. At first the somnambulist could not see it distinctly; he said that the gold setting dazzled him. When the setting was covered with the fingers, he said that he saw the emblem of fidelity. When pressed to tell what this emblem was, he added, ' I see a dog, he is as if on his hind legs before an altar.' This, in fact, was what was represented. A closed letter was presented to him. He could not discover any of its contents. He only followed the direction of the lines with his finger; but he easily read the address, although it contained a pretty difficult name: *A. M. de Kockenstroh.*

"All these experiments were exceedingly fatiguing to M. Petit. He was allowed to repose for an instant. Then, as he was very fond of play, a game at cards was proposed for his relaxation. As much as the experiments of pure curiosity seemed to annoy him, with so much the more ease and dexterity did he perform whatever gave him pleasure, and this he entered into of his own accord. One of the gentlemen present, M. Raynal, formerly inspector of the university, played a game at piquet with M. Petit and lost it. The latter handled his cards with the greatest dexterity, and without making any mistake. We attempted several times in vain to set him at fault, by taking away or changing some of his cards. He counted with surprising facility the points marked upon his adversary's marking card.

"During all this time, we never ceased to examine the eyes, and to hold a candle near them; and we always found them exactly closed. We remarked, however, that the ball of the eye seemed to move under the eyelids, and to follow the different motions of the hands. Finally, M. Bourdois declared, that, according to all human probability, and as far as it was possible to judge by the senses, the eyelids were exactly closed.

"While M. Petit was engaged in a second game at piquet, M. Dupotet, upon the suggestion of M. Ribes, directed his hand, from behind, towards the patient's elbow, and the contraction previously observed again took place. Afterwards, upon the suggestion of M.

Bourdois, he magnetised him from behind, and always at the distance of more than a foot, with the intention of awakening him. The keenness with which the somnambulist was engaged in play, resisted this action, which, without awakening, seemed to annoy and disconcert him. He carried his hand several times to the back of his head, as if he suffered pain in that part. At length he fell into a state of somnolency, which seemed like a slight natural sleep; and some one having spoken to him when in this state, he awoke as if with a start. A few moments afterwards, M. Dupotet, always placed near him, but at a certain distance, set him again asleep, and we recommenced our experiments. M. Dupotet being desirous that not the slightest shadow of doubt should remain with regard to the nature of the physical influence exerted at will upon the somnambulist, proposed to place upon M. Petit as many bandages as we might think proper, and to operate upon him while in this state. In fact, we covered his face down to the nostrils with several neckcloths; we stopped up with gloves the cavities formed by the prominence of the nose, and we covered the whole with a black handkerchief, which descended, in the form of a veil, as far as the neck. The attempts to excite the magnetic susceptibility, by operating at a distance in every way, were then renewed; and, invariably, the same motions were perceived in the parts towards which the hand or the foot were directed.

"After these new experiments, M. Dupotet having taken the bandages off, M. Petit played a game at écarté with him, in order to divert him. He played with the same facility as before, and continued successful. He became so eager at his game, that he remained insensible to the influence of M. Bourdois, who, while he was engaged in play, vainly attempted to operate upon him from behind, and to make him perform a command intimated merely by the will.

"After his game, the somnambulist rose, walked across the room, putting aside the chairs, which he found in his way, and went to sit down apart, in order to take some repose at a distance from the inquisitive experimentalists, who had fatigued him. There M. Dupotet awakened him at the distance of several feet; but it seemed that he was not completely awake, for some moments afterwards he again fell asleep, and it was necessary to make fresh efforts in order to rouse him effectually. When awake, he said he had no recollection of anything that took place during his sleep."

The commissioners also attest other cases of the same kind. I quote, as before, the words of the report.*

"Upon the 12th of January, your committee met again at the house of M. Foissac, where there were present M. E. Lascase, deputy, M. De ——, aide-de-camp to the king, and M. Segalas, member of the Academy. M. Foissac told us that he was going to set Paul asleep, that, in this state of somnambulism, a finger should be applied to each of his closed eyes, and that, in spite of this complete occlusion of the eyelids, he should distinguish the colour of cards, that he should read the title of a work, and even some words or lines pointed out at random in the body of

* This Report of the French Academicians will be found translated at the end of the Isis Revelata, by M. Colquhoun, which I earnestly recommend the English reader to consult.

the work. At the end of two minutes of magnetic manipulations, Paul fell asleep. The eyelids being kept closed, constantly and alternately by MM. Fouquier, Itard, Marc, and the Reporter, there was presented to him a pack of new cards, from which the paper covering bearing the government stamp was torn off. The cards were shuffled, and Paul easily and successively recognised the king of spades, the ace of clubs, the queen of spades, the nine of clubs, the seven of diamonds, the queen of diamonds, and the eight of diamonds.

"While his eyelids were kept closed by M. Segalas, there was presented to him a volume which the Reporter had brought along with him. He read upon the title-page: *Histoire de France.** He could not read the two intermediate lines, and upon the fifth he read only the name, *Anquetil*, which is preceded by the preposition *par*. The book was opened at the 89th page, and he read in the first line—*le nombre de ses*—he passed over the word *troupes*, and continued: *au moment où on le croyait occupé des plaisirs du carnaval.* He also read the running title *Louis*, but could not read the Roman cypher which follows it. A piece of paper was presented to him, upon which were written the words, *Agglutination* and *Magnétisme Animal.* He spelt the first, and pronounced the two others. Finally, the *procès-verbal* of this sitting was presented to him, and he read very distinctly the date, and some words which were more legibly written than the others. In all these experiments, the fingers were

* Histoire de France depuis les Gaulois jusques, à la mort de Louis XVI., par Anquetil. 13 vols. 8vo. Paris, 1817.

The passage read by Paul is to be found upon the 89th page of the 7th volume.

applied to the whole of the commissure of both eyes, by pressing down the upper upon the under eyelid; and we remarked that the ball of the eye was in a constant rotatory motion, and seemed directed towards the object presented to his vision.

"Upon the 2nd of February, Paul was placed in a state of somnambulism in the house of Messrs. Scribe and Bremard, merchants, Rue St.-Honoré. The Reporter of the committee was the only member present at this experiment. The eyelids were closed as before, and Paul read, in the work entitled *Les mille-et-une nuïts*, the title-page, the word *préface*, and the first line of the preface, with the exception of the word *peu*. There was also presented to him a volume entitled, *Lettres de deux amies, par Madame Campan.* He distinguished on a print the figure of Napoleon; he pointed out the boots, and said that he also saw two female figures. He then read currently the four first lines of the third page, with the exception of the word *raviver*. Finally, he recognised, without touching them, four cards, which were successively presented to him two and two,—these were, the king of spades and the eight of hearts, the king and queen of clubs.

"At another sitting, which took place upon the 13th of March following, Paul attempted in vain to distinguish different cards which were applied to the pit of the stomach; but he read, with his eyes still closed, in a book opened at random; and at this time it was M. Jules Cloquet who kept his eyes shut. The reporter also wrote upon a slip of paper the words, *Maximilien Robespierre*, which he read equally well."

Hitherto I have preferred adducing evidence which does not rest altogether on my own authority;

cases operated upon by other physicians, and witnessed by careful and competent observers. It remains for me, however, to state, that I have repeatedly produced the same phenomena. Only a few years ago, I magnetised a young woman who had been, for a distressing complaint, under the treatment of almost all the physicians in Paris. She was attended more particularly by M. Fouquier, but after all was only cured by magnetism in the somnambulic state. Anything intended for her to read was applied to her stomach, taking care to cover the paper with the hand, upon which she told first one letter, then another, and so on throughout the reading. But by an unaccountable *bizarrerie*, she always began to read a word by its termination, and it was necessary to recompose every word in order to understand it; but, notwithstanding, all the letters and words she named were perfectly correct. She said that she felt during the operation the presence of a weight at her stomach, which fatigued her very much. Being questioned how she could see, she could not distinctly state; it seemed to her neither like seeing nor feeling, but like something between the two. One day a snuff-box, perfectly closed, was put into her hands; after having examined it with much attention, she said that it contained in its interior a very small bit of paper. Nobody present was aware of the circumstance; for the box had not been opened since it had been bought, and the gentleman to whom it belonged was by no means prepared for this experiment. But the somnambule, continuing her observations, said that she saw something written on the paper, and she then drew with her finger the figures, of 2 and 5. On opening the box the paper was found, in size one line and a half, and upon it the two small figures which she had described.

A host of additional facts of the same description are in my possession, but the evidence already brought forward will be considered conclusive to every impartial mind Not only does the magnetic somnambulist enjoy, in this state of clairvoyance, a lucidity respecting surrounding objects and passing events, but it extends still further, inasmuch as they foresee, and distinctly foretell, the day, the hour, even the minute, when certain events will happen. This is termed the faculty of prevision, and occurs only in the very highest degree of magnetic exaltation. It may appear incredible; but let us look to the facts. In the treatise on somnambulism by Dr. Bertrand, we find that this physician witnessed upwards of sixty convulsive fits of the most aggravated nature, which it was absolutely impossible the patient could have simulated, all of which, and their duration to a minute, had been foretold during the magnetic sleep. The same physician assures us that a somnambulist told him a fortnight previously, that she should be seized with a delirium which would last forty-three hours, which happened according to her prediction; and he further satisfied himself that this somnambule when awake had not the slightest recollection of what she had said in her sleep. This prevision was also noticed by Drs. Rostan and Georget, one of whom observes:—" I have seen, positively seen, and repeatedly, somnambulists foretell, several hours or days—even twenty days in advance, the precise hour and minute of the accession of epileptic or hysterical fits, and describe what would be their duration and intensity, which came to pass in exact accordance with the prediction." I have myself witnessed innumerable instances of this faculty of prevision being developed during the magnetic somnambulism, but, as my authority would doubtless be impugned as *ex parte* evidence, I prefer again appealing to the report of the French Academicians, the validity of whose testimony cannot reasonably be challenged.

In the case of Paul, already cited, the commissioners attest—" 1st, that he was cured, after every ordinary medicinal resource had failed, of a paralysis, by the treatment being adopted which he himself prescribed during his somnambulism; 2nd, that his strength in the somnambulic state was remarkably increased; 3rd, that he gave us the most undoubted proofs that he read with his eyes closed; 4th, he predicted the period of his cure, and this cure took place.

" In the following case, we shall see this foresight still more fully developed in a man belonging to the lower class, quite ignorant, and who, assuredly, had never heard of animal magnetism.

" Pierre Cazot, aged twenty years, by trade a hatter, born of an epileptic mother, had been subject for ten years to attacks of epilepsy, which occurred five or six times a week, when he was admitted into the Hôpital de la Charité, about the beginning of the month of August, 1827. He was immediately subjected to the magnetic treatment, was set asleep at the third sitting, and became somnambulist at the tenth, which took place upon the 19th of August. It was then, at nine o'clock in the morning, that he announced to us that at four o'clock of the afternoon of that day he should have an attack of epilepsy, but that it might be prevented by magnetising him a little previously. We preferred verifying the exactness of his prediction, and no precaution was taken to prevent its fulfilment. We contented ourselves with observing him, without exciting in him any suspicion. At one o'clock, he was seized with a violent head-ache. At three, he was obliged to go to bed, and precisely at four the fit came on. It lasted five minutes."

On many other occasions Cazot also foretold the time, even to the minute, when his next fit would occur. He was narrowly watched; the Academicians tried to deceive, and prevaricated with him, but in vain, his prediction was always verified. Hence, in the summary of the facts they witnessed, the commissioners observe, towards the conclusion of their report:—

"In two somnambulists we found the faculty of foreseeing the acts of the organism more or less remote, more or less complicated. One of them announced, repeatedly, several months previously, the day, the hour, the minute, of the access, and of the return of epileptic fits. The other announced the period of his cure. Their previsions were realised with remarkable exactness. They appeared to us to apply only to acts or injuries of their organism."

The phenomena of magnetic somnambulism, it will be observed, are thus very similar to those which occur during natural somnambulism. Both states of being are equally mysterious and perplexing.—The Baron Massias draws between them the following diagnosis:—

"Artificial or magnetic somnambulism," says he, "is the result of certain processes, consisting in gestures, manipulations, and acts of volition, operating on persons susceptible of exhibiting its phenomena; it is, as it were, inoculated by the magnetiser. The natural state of the patient submitted to his action when susceptible of receiving it, is generally separated from the subsequent magnetic state by a somnolence of a few minutes. In the crisis a great number of faculties lie dormant, which has caused animal magnetism to retain the name of somnambulism, although magnetism may sometimes be induced without sleep.

" Whereas natural somnambulism is exclusively organic and personal, and places the somnambule in connexion with no one but himself; artificial somnambulism brings the somnambule in magnetic report with the magnetiser, as well as those persons with whom the latter establishes the communication.

" A part only of the intellectual faculties of the somnambule remain awake; they are all so in the magnetic somnambule with regard to those with whom he is in contact *en rapport*. The natural somnambule is impelled by his own imagination, and certain corporeal impressions; the magnetic somnambule wholly depends on the will and faculties of his magnetiser. The organisation of the magnetic somnambule, that of his magnetiser, and of those persons whom the latter brings in report with him by means of a slight contact, are all, as it were, identified as one and the same being; so that the somnambule knows inwardly what takes place in the system of another person; feels diseases which are not his own, and has an intuitive instinct of the remedies applicable to them. This it is which, to the unspeakable scandal and indignation of classic medicine, gave birth to her rival sister, under the name of magnetic medicine. The magnetic somnambule is conscious of the state and organisation of an absent person on merely touching a vestment which has been worn, or an object which has been touched, by the same person. He can either recal or forget, at the will of his magnetiser, what he has felt during the fit of somnambulism. His sensations and perceptions may, at the will of the magnetiser, be changed; or, in other words, he will find sweet that which is bitter, and *vice versa.* He may also change his predilections and habitual dispositions.

"Sometimes, when the action of magnetism overpowers the strength of the somnambule, or through a certain predisposition of his own organs, he collapses into a profound lethargy, bordering upon death.

"It might be thought that the intellectual faculties of the somnambule, plunged in this state of complete lethargy, are as inert and void as his bodily powers: no such thing. There are thoughts, dreams, and somnambulism, in somnambulism. On recovering from his lethargy, and being replaced into his previous state of somnambulism, the crisiac relates all the delightful and marvellous thoughts and sensations he has had during his apparent state of absolute insensibility. What to the spectator was a scene of death, to him was a new life, a hundredfold more intense and spiritual than his ordinary mode of existence. A reiteration of the same fit brings on a recurrence of the same visions. When restored to the state of simple somnambulism, its individual memory returns with it, to be only obliterated by a restoration to the perfect waking state," &c.*

But notwithstanding these observations of the learned Baron, the psychical phenomena exhibited by natural somnambulists are often of so high and lucid a character, that the two states appear almost identical with each other; in illustration of which I shall here quote the case of a girl named Arron, a remark-

* See his Traité de Philosophie Psycho-physiologique. F. Didot. 1830.

able natural somnambulist, which will be found in the Journal de la Meuse for the 20th of September, 1835.

"We readily grant," says the account, "that most of the statements circulated on the subject of natural somnambulism are generally exaggerated. With the view of ministering to the public taste for the marvellous, truth has too often been clothed with most absurd fables, exceeding by far the utmost boundary of our belief. But notwithstanding this, truth is sometimes stranger than fiction; and if the limits of our intelligence do not allow us to comprehend what to us appears supernatural, we should not on that account reject as a mere fabrication any series of facts supported by the testimony of grave and conscientious observers coming forward and vouching for their truth."

"Therefore, however wonderful the details respecting the somnambulism of Arron, we cannot entertain the least doubt of its authenticity, for we ourselves derived it from a source entitled to our utmost and most unlimited confidence."

"This girl, when plunged in a state of somnambulism, answers with precision the questions put to her. Though she be asleep, she perceives not only such external objects as are around her, but also those which are concealed; and, what is still more surprising, objects removed to a very great distance. Nay, she does more, she can divine the thoughts of those who put questions to her. Many physicians in this department went to pay her a visit, and they were all amazed on witnessing a phenomenon which all their science cannot explain."

"A physician from Chartres saw her some time ago. On being introduced to her, in company with several other gentlemen, he questioned her, without being able to obtain an answer. Thinking that if he was alone with her, she might perhaps be induced to speak, he requested the spectators to withdraw. When they were both in private, the following conversation took place :—

" Marie," said he, " do you know me ? "—" Yes, sir." " Who am I ? "—" You are a physician." " Whence do I come ? "—" From Chartres." " Where is my house at Chartres ? "—" In a small street running down a declivity." " Can you see my house ? " —" Yes, sir." " Is there any company in it ? "— " Yes, sir ; four ladies, one old, two middle-aged, and one young lady." " For what purpose have I come in this part of the country ? "—" To see a female patient." " Where is her complaint ? "—(Here she pointed to the part affected, which we cannot just now recollect.) " Where did I dine ? "—" At M.'s." " Was there a good dinner ? "—" Yes, sir." " Could you tell me what dishes we had ? "—" Certainly." (She names every dish and its particular place on the table.) " What do I hold in my hand ? "—" A small wooden box." " What does it contain ?"—" Sharp little iron tools." " Now what have I in my hand ? " —" Some money." " How much ? "—(She names the sum.) " In what coins ? "—(She specifies the various coins.) " Can you tell me my thought at this moment ? "—" Yes, sir." " Say it."—" I dare not ; I must not tell you." " Well, I will tell you : I think of giving you this money."—" So you do, sir ; but I could not say so." All these answers were perfectly correct. Other answers no less surprising than the preceding have been reported to us, but we shall confine ourselves to these.

"Sceptics (and less would suffice to provoke scepticism) will, no doubt, exclaim that it is impossible to credit such assertions; but let them inquire into the evidence, and they will then come to the conviction that we state nothing but the truth."

Hence from the facts above adduced, which I hold to be incontrovertibly established, it will appear that the following are some of the more ostensible psychical phenomena which magnetic somnambulists exhibit.

1st. While the organs of the senses are in a state of complete insensibility, unexcitable by any external *stimuli,* they mentally take cognizance of the conditions and relations of surrounding objects, through some other channel than the organs through which such impressions are usually conveyed.

2nd. There is in many cases an obvious vicarious transference of the senses from their appropriate organs to other parts of the nervous system, as to the tips of the fingers, epigastrium, and other parts of the body.

3rd. The lucidity of their vision penetrates through intervening opaque objects, and even takes cognizance of events passing at a distance.

4th. They possess the faculty of self-intuition, that is to say, a clear insight into the normal or abnormal condition of their own organisation, and they perceive and describe with exactness the internal condition also of those with whom they may be *en rapport.*

5th. They appear endowed with a knowledge beyond that which they ordinarily possess, and prescribe for themselves, as well as for those with whom they are *en rapport,* remedies for such complaints as they may be afflicted with ; and these are generally found successful.

6th. Their lucid vision often extends beyond the present existence, and they foretell, with circumstantial minuteness, events which will happen in connexion with their own organisation ; even the day and hour—the very moment—when such predictions will be verified is accurately specified.

Lastly. On recovering from their somnambulism, they remember nothing which occurred during the magnetic state ; but when again thrown into somnambulism, the memory between the two magnetic states is continuous.

These may appear extraordinary assertions, but the apparent incredibility of the facts will be very much diminished when it is found that analogous, if not identical, phenomena are often developed in abnormal conditions of the human body. To these states I can only briefly advert.

I. Physical insensibility.—In many diseases, in catalepsy, epilepsy, apoplexy, paralysis, syncope, &c., a complete state of insensibility is observed ; but whether persons in such apparent states of insensibility may not possess some internal consciousness, cannot possibly be determined. As a general principle, in proportion as the mind energises itself within, the sensibility of the body becomes diminished. The soldier, on the field of battle, feels not his wounds ; the exulting martyr, in the hour of execution, knows not the death-pang that rends his heart ; the excited mother saves her child by rushing herself through the flames unhurt. There is a mastery here, a triumph of mind over matter which bears down and annihilates the very apprehension of physical suffering. Even the poor maniac, whom madness is supposed to torture in proportion as his mind is excited, manifests insensibility to pain; he may be flagellated while he raves, but feels not the stripes which are inflicted until the paroxysm is over. He then complains of soreness, and is often at a loss to explain how it has occurred. The ordeal of torture to which religious enthusiasts voluntarily subject themselves, without manifesting the slightest sign of pain, proves the insensibility which may be superinduced by the sustained energy of the will. This state is finely described by Schlegel, in his admirable work on the Philosophy of History. He there presents us with a graphic picture of the Indian Yogi, or penitent, who, absorbed in mystic contemplation, remains for years often fixed immovably to a single spot. "In order," says he, "to give a lively representation of a phenomenon so strange to us, which appears totally incredible, and almost impossible, although it has been repeatedly attested by eye-witnesses, and is a well-ascertained historical fact, I shall extract from the Drama of Sacontala, by the Poet Calidas, a description of a Yogi, remarkable for its vivid accuracy, or, to use the expression of the German Commentator, its 'fearful beauty.' King Dushmanta inquires of Indras Charioteer the sacred abode of him whom he seeks, and to this the charioteer replies—'A little beyond the grove you see a pious Yogi, motionless as a pollard, holding his thick bushy hair, and fixing his eyes on the solar orb. Mark! His body is half covered with white ant's edifices, made of raised clay; the skin of a snake supplies the place of the sacerdotal thread, and part of it girds his loins; a number of knotty plants encircle and wound his neck, and surrounding birds' nests conceal his shoulders.'" "We must not," continues Schlegel, "take this for the invention of fancy, or the exaggeration of the poet; the accuracy of this testimony is confirmed by innumerable eye-witnesses, who recount the same fact, and in precisely the same colours. In the Indian forests and deserts there are many hundreds of these hermits,—these strange human phenomena of the highest intellectual abstraction and delusion. Formerly such accounts would have been regarded as incredible, and beyond the bounds of possibility; but such conjectures can be of no avail against historical facts, repeatedly attested, and undeniably proved. Now that men are better acquainted with the wonderful flexibility of the human organisation, and with those marvellous powers which slumber concealed within it, they are less disposed to form light and hasty decisions on phenomena of this description. The whole is indeed a magical, intellectual, self-exaltation, accomplished by the energy of the will; and this concentration of the mind, when carried to excess, may lead, not merely to a figurative, but to a real self-annihilation."*

The state of intellectual *exstase*, accompanied by an absolute extinction of physical sensibility, induced by animal magnetism, is precisely analogous to this mysterious state of being. Here also I may observe that the *criteria* of this profound insensibility demand attention, inasmuch as it is quite true that needles may, as in the operation of acupuncturation,

* Philosophy of History, by Frederick Von Schlegel. Translated by James Burton Robertson. 2 vols. London, 1837.

be plunged into the muscles of a limb without exciting pain; that the sensibility of some persons with a slow circulation, or leuco-phlegmatic habit, may be often remarkably obtuse; that some may be sensible of pain, and yet command sufficient energy of mind to suppress any outward manifestation of it. Yet, notwithstanding all this, when we find that a glare of light thrown suddenly upon the eye excites no contraction of the pupil; that fuming ammonia introduced into the nasal passages does not excite sneezing; that the report of a pistol unexpectedly fired close upon the ear does not induce even a shudder, or any appearance of sensation; it is impossible to resist the evidence of there being in these organs a complete extinction of all physical sensibility.

This state of insensibility, accompanied by somnambulic phenomena, sometimes arises spontaneously, and without any ostensible magnetic intervention. "I lately," says Dr. Abercrombie, the present eminent physician in Edinburgh, had "under my care a lady who is liable to an affection of this kind (somnambulism), which comes on repeatedly during the day, and continues from ten minutes to an hour at a time. Without any warning, her body becomes motionless, her eyes open, fixed, and entirely insensible, and she becomes totally unconscious of any external impression. She has been frequently seized while playing on the piano, and has continued to play over and over a part of a tune with perfect correctness, but without advancing beyond a certain point. On one occasion she was seized after she had begun to play from the book a piece of music which was new to her. During the paroxysm she continued the part she had played, and repeated it five or six times with perfect correctness; but on coming out of the attack, she could not play it without her book.*

II. Clairvoyance, or vision without the eyes.—In the most simple cases of natural somnambulism this phenomenon occurs. The person who rises in the night, dresses himself, moves from room to room, and performs with nicety and precision a variety of minute actions, presents us with this anomaly in its most familiar form. There are few sceptics who do not acknowledge cases of this kind; for after all, somnambulism is a very common occurrence, and is observed to be hereditary in many families. Yet when this faculty becomes more perfectly developed, manifesting itself by finer discriminations, the very same philosophers, forgetting the fact which is universally accredited in domestic life, open their eyes with astonishment, and declare the whole thing utterly incredible. This is unreasonable. It is a notorious fact, that all sleep-walkers are clairvoyants.

It has been said, however, that this mode of perception cannot exactly be called seeing, inasmuch as when such somnambulists are closely examined how it is that they perceive the relations of external objects, they all agree in answering that it is by a kind of inward feeling which they are unable to explain. A somnambulist of Fischer assured him that he saw his internal organs, not as with the eyes, but he could not describe the manner in which he perceived them. A patient of Hufeland's said, "*I see*," when she was in the highest degree of lucidity only, at other times she generally used the expression "*I feel*," this or that

* Inquiries concerning the Intellectual Powers and the Investigation of Truth, by John Abercrombie, M.D. Edinburgh, 1832.

part, or this or that change, &c. A patient of Gmelin's said she did not *see*, but feel, and with great delicacy, both internally and externally. In the magnetic sleep, Scherb's patient also declared that her sensations were rather those of *feeling* than of *sight*, and that this peculiar feeling during that state was much more acute and delicate than when awake. A cataleptic patient of Despine's said to those who were *en rapport* with her, "You think that I do not know what passes around me, but you are mistaken; I *see* nothing, yet I *feel* something which makes an impression upon me, but which I cannot explain."* These are curious facts. How far the sense of feeling may, under a peculiar state of exaltation, take cognizance of the outward forms and conditions of material objects remains yet to be determined; certain it is, that in well-authenticated cases, both of natural and magnetic somnambulism, individuals, through some other channel than the eye itself, perceive, discriminate, and perform a variety of actions which could not be accomplished without vision being in some way or other engaged.

In the animal economy it is a physiological law

* Vide Isis Revelata, loc cit. Her words were, "On croit que je ne sais pas ce qui se passe autour de moi tous les soirs; mais on se trompe bien. Je ne vois rien, je n'entends rien, mais je *sens* quelque chose que je ne puis m'expliquer, qui me maîtrise."

The fact of the sense of feeling substituting itself, under certain conditions, for that of sight, has not altogether escaped the attention of poets as well as men of science. Thus the philosophical author of the Cosmo de Medici, R. H. Horne, in his tragedy, the Death of Marlowe, makes one of his heroes observe—

> Oh, subtle Nature! who hath so confounded
> Our senses, playing into each other's wheels,
> That *feeling* oft acts substitute for *sight*,
> *   *   *   I *feel*—I see.

that one organ shall assume a preternatural activity in order to supply the function of another whose structure may be deranged; thus it is clearly proved that the liver acts vicariously with the lungs; the skin with the kidneys. It is the office of the lungs to throw out of the system, during expiration, carbonaceous matter; and when, therefore, this organ is diseased, the liver assumes, in a manner, its function, and throws off the surplus carbon with the biliary secretion. Hence the frequency of diseases of the liver in those persons who have resided in tropical countries, where, owing to the climate, the lungs may have become affected. In like manner the vessels of the skin, acting vicariously with those of the kidneys, throw off the watery parts of the blood when the secretion of this organ has, from any cause, been obstructed. Here there is an unquestionable transference of function between the grosser organs of organic life—and why should not the higher organs of sense, under certain abnormal conditions, act on a similar principle, vicariously with each other? It is well known that persons who unhappily are blind and deaf acquire a preternaturally acute sense of touch, so that they are able to discriminate by feeling, properties of bodies which could not be so appreciated by other persons. This, however, is only a very subordinate physical manifestation of that law of derivation to which I here allude; it may, nevertheless, be the initiatory condition of that transference of the senses which is in the magnetic state much more perfected.

III. Transference of the senses.—The clairvoyant faculty sometimes develops itself locally, requiring that the object to be perceived shall be placed in juxta-position with certain parts of the body, as with the epigastrium, the tips of the fingers, the forehead, or even back part of the head. It is not during the magnetic somnambulism only that this extraordinary mode of perception develops itself; it occurs often in catalepsy and hysterical affections. A cataleptic patient of Petetin distinguished in succession several cards which had been slipped under the bed-clothes, and laid upon her stomach; she told the hour of a watch held in the closed hand of an inquirer; and recognised, in like manner, an ancient medal grasped in the hand of one of the spectators. Another day the same patient recognised a letter placed under the waistcoat of M. Petetin, then a purse which had been slipped there by a sceptic; she also indicated the number of gold and silver coins contained in each end of the purse. After this experiment, she announced that she was about to tell successively what each person present possessed about him most remarkable, which she actually did. She also perceived, through a screen, that Madame Petetin on going out took her husband's cloak, and made him discover the mistake.

Jourdain Guibelet relates the case of a young lady labouring under hysteria, who, during her fits, which lasted upwards of twenty-four hours, lay to all appearance in a state of absolute insensibility; yet were the organs of speech under no impediment; she discoursed with so much judgment and refined wit, that it seemed as if her disease had sharpened her intellect, and been more liberal to her than health itself. No one could have reasoned with so much precision, or discoursed with so much fluency. It might be said that the body being as it were dead, under the violence of the disease, the soul retired into her chamber, where she fully enjoyed all her privileges. The conceptions of the soul in that state are the more accurate and refined in proportion as she is more free from the trammels of the body and of matter.

Bertrand relates the case of a lady, Mademoiselle A., who exhibited a remarkable singularity as to the way in which she received sensations either by the pit of the stomach, fingers, or toes. When an odorous substance was applied to the tips of the fingers, she cried "Ah que cela sent bon!" and at the same time she inhaled the air with her nostrils. It was the same in respect to sight; when an object was placed on the palm of the hand, the soles of the feet, or the epigastrium, if she said, "Je ne vois pas bien," she rubbed her eyes with her hand, as if awake.*

IV. Exaltation of knowledge.—That persons during magnetic somnambulism possess a knowledge beyond that which they possess in their waking state, is also certain. It is a repetition of the same phenomenon which occurs in natural somnambulism; furthermore, it has been observed that on the accession of certain diseases, especially fever attended with head symptoms, the intellectual faculties are preternaturally acute. This is an old observation of physicians. It is stated by Levinus Lemnius, (lib. 2, cap. 2, Collect. de Occult Nat. Miraculis), that he himself healed some sick persons who in their fits of fever pronounced an oration as if it were deeply studied, and were in all respects most accomplished, yet who in health were little better than idiots. Cardan (De Rerum Variet., lib. 8, cap. 43), relates that Phliarius Poletanus, an

* Bertrand, du Magnétisme Animal en France. Paris, 1826.

Italian, distempered in body, spoke the Dutch tongue, which he had never learned; and that after passing several worms, and recovering, he lost that ability, and could then only speak his own language. Petrus Apponensis (Comment, at Problem 1, sect. 3, de Aristot.), mentions a woman who in a melancholy sickness spoke Latin; yet on recovering could not speak a word of it. Sigebertus Continuator says, that Norbert of Nigella, when grievously sick, did, from the devil, repeat the Canticles from one end to the other. A host of authorities might be cited to the same effect.

Bertrand mentions two female somnambulists who expressed themselves, during the paroxysm, very distinctly in Latin; they had in their waking state, however, only a very imperfect knowledge of it. An ignorant servant girl, mentioned by Dr. Dewar, during paroxysms of this kind shewed an astonishing knowledge of geography and astronomy; and expressed herself in her own language in a manner which, though often ludicrous, shewed an understanding of the subject. The alternations of the seasons, for example, she explained, by saying that the earth was set *agee;* it was afterwards discovered that her notions on these subjects had been derived from overhearing a tutor giving instructions to the young people of the family. A woman who was some time ago in the infirmary of Edinburgh, on account of an affection of this kind, during the paroxysms mimicked the manner of the physicians, and repeated correctly some of their prescriptions in Latin.*

V. Prevision.—This, which is the highest degree of

* Abercrombie's Intellectual Powers, *loc. cit.*

clairvoyance, has been occasionally observed by physicians to occur on approaching death. "In dying," says Dr. Stellung, "the person loses his consciousness; he falls into a perfect trance—a profound sleep. As long as the mass of the blood is warm, and not congealed, all the members of the body continue pliant, and as long as this is the case the soul remains in it; but as soon as the brain and nerves lose their warmth and become frigid, they can no longer attract the etherial part of the soul, or retain it any longer; it therefore disengages itself, divests itself of its earthly bonds, and awakes. It is now in the state of a clear-seeing magnetic sleeper, and being entirely separated from the body, its state is much more perfect; it has a complete recollection of its earthly state from beginning to end; it remembers those it has left behind, and can form to itself a very clear idea of the visible world and its objects; viz., that part of it to which it belongs, or to which it has here adapted itself. The candid inquirer will easily find that all this follows logically and justly from magnetic experiments, if he be acquainted with them, and duly consider them."

Aretœus, in his chapter on the Causes of Ardent Fever, notices this faculty in the following terms:—"The mind is serene, every sense clear, the understanding subtle, the judgment prophetic; for the patients see, in the first place, that their end is approaching, and then they predict to the bystanders coming events. Sometimes people think them talking deliriously, but when their predictions come to pass, are amazed. Some hold converse with the dead, whom they alone perceive as present, from a subtle and fine perception, or perhaps from a prescient spirit which announces to them the men in whose company they will shortly be; for though previously in the thick of humours and darkness, the disease having drawn them out and cleared the mist from their eyes, they describe these airy visions, and with a spirit unencumbered (leur ame étant, pour ainsi dire, mise à nud), become infallibly prophetic. They who have arrived at this pitch of subtlety of the humours and understanding, rarely recover, for the vital power is already melting into air."*

Lucidity and prevision, occurring as symptomatic of disease, were noticed by Hippocrates, who therefore called melancholia, monomania, phrensy, and certain other diseases of the same class, *divine*, because "in them," said he, "men without any learning speak Latin, make verses, discourse with eloquence and wisdom, predict secret and future events; and this they do, albeit, they are ignorant rustics, or even idiots; and after being cured they return to their former state." "In ecstacy," says Aristotle, "the mind darts forward through the relations of causes and effects, seizes the whole concatenation with the quickest glance, and submits it to the imagination in order to infer from it the probability of future events. Melancholy persons, on account of the vehemence of their imagination, are generally fittest for such conjectural operations." Sauvage, in his Nosologie, quotes a passage of Descartes,

* Aretœus on the Causes and Signs of Acute and Chronic Diseases, translated from the Greek, by T. F. Reynolds. London: Pickering, 1837. Or, the French Translation, par M. L. Renard. Paris, 1834.

"Divinare autem morientes eliam illo exemplo confirmavit Possidonius quo affert. Rhoduum quemdam morientem sex equales nominasse, qui primus eorum, qui secundus, qui deinceps moriturus esset."—Cicero de Divinatione, lib. 1.

in which he relates that two hysterical young females used to foretel to each other the various crises of their respective disease. He also refers to Cavaliers, who states that he saw, in Frejus, "four hydrophobic patients, who had foretold the day, and even the hour, of their death, and who died on the very day and hour predicted. I, myself," says he, "saw a sexagenarian predicting the day of his death one month in advance; and he died of an epial fever on the day predicated." Other facts of the same description have been attested by Deseze; and recently by Professor Moreau de la Sarthe, in an article headed Médecine Mentale.

During ordinary sleep, or natural somnambulism, this same faculty often developes itself; hence it is well observed by Hippocrates, that "he who could obtain a perfect knowledge of the various conjectures that may be inferred from dreams, might reap from it considerable benefit in his intercourse with mankind. When the body is at rest, the soul begins to move of her own accord, and, receding gradually from the different parts of the body, retires and concentrates herself in her own seat. Here she performs all the functions of the body; for the body, when overcome by sleep, feels nothing; it is then the soul that wakes, knows, sees what is to be seen, hears what is to be heard, feels, grieves, and reasons more readily, and with greater facility. During sleep she performs all the functions, her own as well as those of the body. If, therefore, any one could comprehend with a sound judgment this peculiar state of the soul in sleep, he might flatter himself of having made an immense stride in the knowledge of wisdom."

"We should not wonder," says Plutarch, "if the soul which can grasp that which is no more, is also able to foresee that which lies unborn in futurity. The future is even much more interesting and congenial to the soul; she longs after the future, and is already grasping it at the point of transition from the past, to which she is connected only by retrospection. Souls, therefore, have this innate faculty, but it is feeble and obscure, and acts but with difficulty. However, in some individuals, she suddenly unfolds her powers, either in dreams, or when the body happens to be in a favourable position for enthusiasm, and when the rational and contemplative part of its essence, freed from the impression of present scenes by which her action is disturbed, directs the imagination towards the prevision of the future." He then adds, "Men when they are awake have but one and the same world common to them all; but in their sleep, each has a private world of his own." "During sleep," observes Tertullian, "is revealed to us, not only what pertains to honour and riches, but likewise what pertains to diseases, remedies, and cures." So also Jamblichus, who, in his chapter on dreams, remarks,—"When we are in a perfect state of sleep, we cannot observe so clearly and distinctly what occurs to us in our common dreams, as when the Deity sends them to us in a sort of unconscious sleep; for then we perceive much more clearly the truth and reality of those things which we have been accustomed to do when awake; by which means such visions constitute the principal species of divination. And, in fact, the soul has then a double existence; the one connected with the body, and the other abstracted from all corporeality. When we awake, we spend most of our time in that life which is inherent in the body, except sometimes when we are totally separated therefrom; but in sleep, our mind may be entirely freed from the corporeal fetters, which keep it, as it were, a prisoner."

Hence this faculty of prevision has been in all ages observed, and is attested by a concurrence of evidence which is irresistible. Whether it arise, as these philosophers conjectured, from any exaltation, or unfolding of the soul, or psychical power which animates the breast of every human being; or whether it proceed from the nervous fluid, is of little consequence in respect to the result; but it appears to me evident, that in every case it is owing to an accumulation of the magnetic fluid in the brain,—in the concatenation of the nerves of the solar plexus;* or in some other part of the nervous system, and this may occur naturally during the progress of certain diseases, as mania, fever, hysteria, &c., or during sleep, or it may be produced by magnetisation. However the cause may vary, a similar pathognomonic state being induced, the effects are identical.

Lastly. The continuity of memory between the paroxysms.—This anomaly often occurs in disease. It is thus noticed by Dr. Abercrombie, whose acute powers of observation and veracity are universally esteemed:—

* It may be proper to apprise unprofessional readers of what is meant by the solar plexus. The nervous system of vertebrate animals is composed of the brain, the spinal cord, and nerves proceeding from, or, more properly speaking, in connexion with them. But besides these there is a system of nerves called the sympathetic, which supply the interior organs of the body. The interlacing, or great concatenation, of these in the neighbourhood of the stomach, or epigastrium, is the solar plexus here referred to. It should be added, that the functions of these nerves are not understood, and magnetisers have observed, that the effects of animal magnetism are more speedily marked in this than in any other region of the body.

"Another very singular phenomenon presented by some instances of this affection, is what has been called, rather incorrectly, a state of double consciousness. It consists in the individual recollecting, during a paroxysm, circumstances which occurred in a former attack, though there was no remembrance of them during the interval. This, as well as various other phenomena connected with the affection, is strikingly illustrated in a case described by Dr. Dyce, of Aberdeen, in the Edinburgh Philosophical Transactions. The patient was a servant girl, and the affection began with fits of somnolency, which came upon her suddenly during the day, and from which she could at first be roused by shaking, or by being taken out into the open air. She soon began to talk a great deal during the attacks regarding things which seemed to be passing before her as a dream; and she was not, at this time, sensible of anything that was said to her. ....In her subsequent paroxysms, she began to understand what was said to her, and to answer with a considerable degree of consistency," &c. "She also became capable of following her usual employments during the paroxysm; at one time she laid out the table correctly for breakfast, and repeatedly dressed herself and the children of the family, her eyes remaining shut the whole time. The remarkable circumstance was now discovered, that, during the paroxysm, she had a distinct recollection of what took place in former paroxysms, though she had no remembrance of it during the intervals. At one time, she was taken to church while under the attack, and there behaved with propriety, evidently attending to the preacher; and she was at one time so much affected as to shed tears. In the interval, she had no recollection of having been at church ; but, in the next paroxysm, she gave a most distinct account of the sermon, and mentioned particularly the part of it by which she had been so much affected."—"During the attack, her eye-lids were generally half shut; her eyes sometimes resembled those of a person affected with amaurosis,—that is, with a dilated and insensible state of the pupil; but sometimes they were quite natural."—"At one time, during the attack, she read distinctly a portion of a book which was presented to her: and she often sung, both sacred and common pieces, incomparably better, Dr. Dyce affirms, than she could do in the waking state."*

Dr. Abercrombie also relates the following analogous history. "A girl aged seven years, an orphan of the lowest rank, residing in the house of a farmer, by whom she was employed in tending cattle, was accustomed to sleep in an apartment separated by a very thin partition from one which was frequently occupied by an itinerant fiddler. This person was a musician of very considerable skill, and often spent a part of the night in performing pieces of a refined description; but his performance was not taken notice of by the child except as a disagreeable noise. After a residence of six months in this family, she fell into bad health, and was removed to the house of a benevolent lady, where, on her recovery, after a protracted

* Abercrombie, On the Intellectual Powers. Fourth Edition, pp. 294, &c.

It appears that this girl was afterwards abused, in one of her paroxysms, in the most brutal and treacherous manner. On awaking, she had no consciousness whatever of the outrage; but in a subsequent paroxysm, some days afterwards, it recurred to her recollection, and she then related to her mother all the revolting particulars.

illness, she was employed as a servant. Some years after she came to reside with this lady, the most beautiful music was often heard in the house during the night, which excited no small interest and wonder in the family; and many a waking hour was spent in endeavours to discover the invisible minstrel. At length, the sound was traced to the sleeping-room of the girl, who was found fast asleep, but uttering from her lips a sound exactly resembling the sweetest sounds of a small violin. On further observation it was found, that, after being about two hours in bed, she became restless and began to mutter to herself;—she then uttered sounds precisely resembling the tuning of a violin, and at length, after some prelude, dashed off into elaborate pieces of music, which she performed in a clear and accurate manner, and with a sound exactly resembling the most delicate modulations of that instrument. During the performance she sometimes stopped, made the sound of re-tuning her instrument, and then began exactly where she had stopped in the most correct manner."

"After a year or two, her music was not confined to the imitation of the violin, but was often exchanged for that of a piano of a very old description, which she was accustomed to hear in the house where she now lived; and she then also began to sing, imitating exactly the voices of several ladies of the family. In another year from this time, she began to talk a great deal in her sleep, in which she seemed to fancy herself instructing a younger companion. She often descanted with the utmost fluency and correctness on a variety of topics, both political and religious, the news of the day, the historical parts of scripture, public characters, and particularly the characters of members of the family and their visitors. In these discussions she shewed the most wonderful discrimination, often combined with sarcasm, and astonishing powers of mimicry. Her language through the whole was fluent and correct, and her illustrations often forcible and even eloquent. She was fond of illustrating her subjects by what she called a fable, and in these her imagery was both appropriate and elegant. She was by no means, says my informant, limited in her range,—Bonaparte, Wellington, Blucher, and all the kings of the earth, figured among the phantasmagoria of her brain; and all were animadverted upon with such freedom from restraint, as often made me think poor Nancy had been transported into Madame Genlis' Palace of Truth. The justness and truth of her remarks on all subjects, excited the utmost astonishment in those who were acquainted with her limited means of acquiring information. She has been known to conjugate correctly Latin verbs, which she had probably heard in the school-room of the family; and she was once heard to speak several sentences very correctly in French,—at the same time stating that she heard them from a foreign gentleman, whom she had met accidently in a shop. Being questioned on this subject when awake, she remembered having seen the gentleman, but could not repeat a word of what he said. During her paroxysms, it was almost impossible to awake her, and when her eyelids were raised, and a candle brought near the eye, the pupil seemed insensible to the light. For several years, she was, during the paroxysms, entirely unconscious of the presence of other persons; but, about the age of sixteen, she began to observe those who were in the apartment, and she could tell correctly their numbers, though the utmost care was taken to have the room darkened. She now also became capable of answering questions that were put to her, and of noticing remarks made in her presence; and with regard to both, she shewed astonishing acuteness. Her observations, indeed, were often of such a nature, and corresponded so accurately with characters and events, that by the country-people she was believed to be endowed with supernatural powers.

"During the whole period of this remarkable affection, which seems to have gone on for ten or eleven years, she was, when awake, a dull, awkward girl, very slow in receiving any kind of instruction, though much care was bestowed upon her; and, in point of intellect, she was much inferior to the other servants of the family. In particular, she shewed no kind of turn for music. She did not appear to have any recollection of what had passed during her sleep; but, during her nocturnal ramblings, she was more than once heard to lament her infirmity of speaking in her sleep, adding, how fortunate it was that she did not sleep among the other servants, as they teased her enough about it as it was."

Here, also, I may adduce the following interesting case, which occurred in a small town in Germany, and is thus related by the late S. T. Coleridge:—"A young woman, of four or five and twenty, who could neither read nor write, was seized with a nervous fever, during which, according to the asseverations of all the priests and monks of the neighbourhood, she became possessed, and as it appeared, by a very learned devil. She continued incessantly talking Latin, Greek, and Hebrew, in very pompous tones, and with most distinct enunciation. This possession was rendered more probable by the known fact that she was or had been an heretic. Voltaire humorously advises the Devil to decline all acquaintance with medical men; and it would have been more to his reputation if he had taken this advice in the present instance. The case had attracted the particular attention of a young physician, and by his statement many eminent physiologists and psychologists visited the town and cross-examined the case on the spot. Sheets full of her ravings were taken down from her own mouth, and were found to consist of sentences coherent, intelligible each for itself, but with little or no connexion with each other. Of the Hebrew a small portion only could be traced to the Bible; the remainder seemed to be in the Rabbinical dialect. All trick or conspiracy was out of the question. Not only had the young woman ever been an harmless simple creature; but she was evidently labouring under a nervous fever. In the town in which she had been resident for many years as a servant in different families, no solution presented itself. The young physician, however, determined to trace her past life step by step; for the patient herself was incapable of returning a rational answer. He at length succeeded in discovering the place where her parents lived; travelled thither, found them dead, but an uncle surviving, and from him learned that the patient had been charitably taken by an old protestant pastor at nine years old, and had remained with him some years, even until the old man's death. Of this pastor the uncle knew nothing, but that he was a very good man. With great difficulty, and after much search, our young medical philosopher discovered a niece of the pastor's, who had lived with him as his housekeeper, and had inherited his effects. She remembered the girl related that her venerable uncle had been too indulgent, and could not bear to hear the girl scolded; that she was willing to have kept her, but that after her patron's death she herself refused to stay. Anxious inquiries were then, of course, made concerning the pastor's habits, and the solution of the phenomenon was soon obtained. For it appeared that it had been the old man's custom for years to walk up and down a passage of his house into which the kitchen door opened, and to read to himself with a loud voice out of his favourite books. A considerable number of these were still in the niece's possession. She added that he was a very learned man, and a great Hebraist. Among the books were found a collection of Rabbinical writings, together with several of the Greek and Latin fathers; and the physician succeeded in identifying so many passages with those taken down at the young woman's bedside, that no doubt could remain in any rational mind concerning the true origin of the impressions made on her nervous system.

"This authenticated case," continues Coleridge, "furnishes both proof and instance, that reliques of sensation may exist for an indefinite time in a latent state in the very same order in which they were originally impressed; and as we cannot rationally suppose the feverish state of the brain to act in any other way than as a stimulus, this fact (and it would not be difficult to adduce several of the same kind) contributes to make it even probable that all thoughts are in themselves imperishable; and that if the intelligent faculty should be rendered more comprehensive, it would require only a different and apportioned organisation, the body celestial instead of the body terrestrial, to bring before every human soul the collective experience of its whole past experience. And this perchance is the dread book of judgment, in whose mysterious hieroglyphics every idle word is recorded;* yea, in the very nature of a living spirit it may be more possible that heaven and earth should pass away, than a single thought should be loosened or

* This same idea is thus beautifully expressed by Mr. De Quincy, in his "Opium Eater;" where, in describing his dreams, he observes:—The minutest incidents of childhood, or forgotten scenes of later years, were often revived: I could not be said to recollect them; for if I had been told of them when waking, I should not have been able to acknowledge them as parts of my past experience. But placed as they were before me, in dreams like intuitions, and clothed in all their evanescent circumstances and accompanying feelings, I *recognised* them instantaneously. I was once told by a near relative of mine, that having in her childhood fallen into a river, and being on the very verge of death but for the critical assistance which reached her, she saw in a moment her whole life, in its minutest incidents, arrayed before her simultaneously as in a mirror; and she had a faculty developed as suddenly for comprehending the whole and every part. This, from some opium experiences of mine, I can believe; I have, indeed, seen the same thing asserted twice in modern books, and accompanied by a remark which I am convinced is true; viz., that the dread book of account, which the Scriptures speak of, is, in fact, the mind itself of each individual. Of this, at least, I feel assured, that there is no such thing as *forgetting* possible to the mind; a thousand accidents may and will interpose a veil between our present consciousness and the secret inscriptions on the mind; accidents of the same sort will also rend away this veil; but alike, whether veiled or unveiled, the inscription remains for ever; just as the stars seem to withdraw before the common light of day, whereas, in fact, we all know that it is the light which is drawn over them as a veil—and that they are waiting to be revealed, when the obscuring daylight shall have withdrawn.

lost from that living chain of causes to all whose links, conscious or unconscious, the free-will our only absolute itself is co-extensive and co-present. But not now dare I longer discourse of this, waiting for a loftier mood and a nobler subject, warned from within and from without that it is profanation to speak of these mysteries."*

From the facts which I have now adduced, it is evident that, however extraordinary they may appear, phenomena are developed in certain abnormal conditions of the body which are similar to those that occur during magnetic somnambulism, and this circumstance alone should suspend the summary judgment of those who are inclined to disbelieve in the possibility of such facts ever occurring; for it is not more marvellous that an utter extinction of physical sensibility, somniloquism, clairvoyance, prevision, &c. should appear during magnetic than during natural somnambulism. Both states are extraordinary, but neither incredible; and after all, it should be remembered that life, mind, the soul, our existence in all its relations, is a mystery, in the very shadow and perplexity of which it is our duty to watch with attention and diligence every fact which, when well attested, forms a part of that induction which may eventually relieve us from the embarrassment of our ignorance. The principle upon which we estimate that which is possible or impossible is erroneous; the experience of the past ought not to impose boundaries on the experience of the future. Our eyes are daily closed to an infinite number of sublime truths, which revolve around us in the simplest occurrences of nature; and

* Biographia Literaria, or Biographical Sketches of my Literary Life and Opinions, by S. T. Coleridge. London, 1817.

he who would unseal them ought not to be repudiated as a visionary or impostor. We forget that every grain of sand on the sea shore has its position determined by laws as immutable as those which maintain the heavenly bodies in their orbits; the one may appear an insignificant, the other a sublime object of contemplation; but the wonder and mystery of the one is not greater than the wonder and mystery of the other. Our philosophers restrict their attention too much to the material, to the total neglect of the immaterial, portion of their being; they forget that mind must be governed by laws as fixed and constant as those which govern the physical universe. "All the organs of sense," says Coleridge, "are framed for a corresponding world of sense, and we have it. All the organs of spirit are framed for a corresponding world of spirit, though the latter organs are not developed in all alike."* This is perfectly true, and I am persuaded that the study of animal magnetism is alone the path which will lead us to any just conception of the laws of our spiritual existence. Let the facts of animal magnetism be sifted and weighed with as much suspicion and care as any other evidence which might be brought into a court of justice; they will, I am satisfied, be thoroughly substantiated, and it will then be found that the consequences to which they lead are of the highest and most sacred importance to the moral government and happiness of mankind.†

* Biographic Lit. *loc. cit.*

† The study of animal magnetism gives a death blow to materialism. The accomplished Dr. Georget acknowledges that it converted him from being a materialist to believe in the immortality of the soul.

CHAPTER IV.

METHOD OF CONDUCTING THE MAGNETIC OPERATION.

THE method of inducing the phenomena of animal magnetism may easily be acquired; there is no mystic art or secresy in the operation; but it should be conducted, for reasons which I shall hereafter explain, with great caution, and strictly upon the principles which the magnetisers have laid down. In the infancy of the science, ceremonies, processes, and precautions were observed, which are now regarded as unnecessary; but in order that the subject may be fairly comprehended, I conceive it right to return to the time of Mesmer, and describe the method he adopted. It should, however, be premised that he believed that the magnetic phenomena were produced by the intervention of agents which we now consider entirely foreign to their causation; that he never mentioned the will as being essential to the production of the results; and that the most curious phenomenon resulting from magnetisation, somnambulism itself, was entirely unknown to him, although every circumstance would lead to the inference that he was not altogether unacquainted with it.

The following is a description of the magnetic process and apparatus which he employed, and which were submitted to the examination of the commissioners in 1784:—In the middle of a large room was placed a circular vessel, termed the *baquet*, a few feet in height, furnished with a lid in two parts, moving on hinges in a central line. This lid was perforated with holes, through which were inserted a number of curved and moveable iron rods. Its interior was filled with bottles full of water previously magnetised. These were placed over one another in such a manner that the first row had their necks converging towards the centre of the vessel, and their bases turned towards the circumference; and the next set was arranged in an opposite position. The baquet itself contained also a certain quantity of water filling up the interstices which were left by this symmetrical arrangement of the bottles; and to this a quantity of iron filings, pounded glass, sulphur, manganese, and a variety of other substances, was occasionally added. The patients then stood round this apparatus, and applied the iron rods to the affected parts of the body, or encircled themselves with a hoop suspended for that purpose. Sometimes they laid hold of each other by the thumb and index finger, and formed what was called a chain. The magnetiser then held an iron rod, which he moved to and fro before them, for the purpose of directing at will the course of the magnetic fluid. The whole apparatus of water, bottles, and metallic rods was supposed to facilitate the circulation of the fluid; and during this time a person occasionally played on the piano or harmonicon; for it was one of Mesmer's axioms, that the magnetic fluid was especially propagated by sound. These processes, which may appear whimsical, formed the basis of the treatment in common, or by the *baquet;* but it being presumed that the universal fluid was every where, the magnetiser himself was thought to possess a certain quantity which he could communicate and direct, either by means of a rod, or simply by the motion of his outstretched fingers. To these gestures, performed at a distance, were also added certain slight touchings on the hypochondria, the epigastric region, or the limbs. In order to increase the power of these processes, trees, water, food, and other objects, were magnetised, for all the bodies in nature are, according to Mesmer, susceptible of magnetisation.

His theory had something seductive in its application. "A non-magnetic needle," said he, "when set in motion, will only take by chance a given direction; whereas, on the contrary, the real magnetic needle, having received the same impulse, will, after a number of oscillations proportionate to that impulse, and to the degree of magnetism it has received, resume its former position, in which it will remain fixed. Thus it is that the harmony of organised bodies, being once disturbed, must undergo such casualties, unless restored and determined by the general agent, the existence of which I admit, and which alone can re-establish the harmony of the natural functions. We also know that, in all ages, diseases have been aggravated or cured with or without the aid of medicine, according to various systems and methods, in direct opposition to each other. These considerations do not allow me to doubt that there is in nature an universal active principle, producing many phenomena, which we vaguely attribute to art and to nature."

The following directions, however, by Mesmer himself, will give a correct idea of the manner in which he wished the operation to be conducted. I quote his own words, from the instructions, which he gave in confidence to his pupils. They are arranged in the form of a catechism :

" Q.—How are the effects of the animal fluid demonstrated ?

" A.—When a healthy person is brought into immediate contact with a sick person, in whom one or more functions are disordered, the latter feels, in the morbid part, sensations more or less acute.

" Q.—How must a patient be touched to make him feel the effects of magnetism ?

" A.—You must place yourself opposite to him, with your back turned towards the north, and draw your own close against his feet; you must then place, without pressure, both your thumbs on the plexus of the nerves in the epigastrium, and stretch your fingers towards the hypochondria. It is beneficial occasionally to move your fingers on the sides, and especially in the region of the spleen. After having continued this exercise for about a quarter of an hour, you should change your mode of operating, according to the state of the patient.

" Q.—What ought to be done before we cease magnetising ?

" A.—You must endeavour to put the magnetic fluid in equilibrium in every part of the body. This may be done by presenting the index finger of the right hand at the summit of the head or the left side, and then drawing it down the face to the breast and over the lower extremities. In this manœuvre an iron rod may be used instead of the finger.

" Q.—Can we not augment the force, or the quantity, of the magnetic fluid on individuals ?

" A.—The power of magnetism is augmented by

establishing a direct communication between several persons.

"Q.—How is this communication established?

"A.—In two ways; the more simple is to form a chain, with a certain number of persons made to hold each other's hands; it can also be done by means of the baquet," &c.

The effects produced by such processes were not less strange than the processes themselves. I have already described certain extraordinary phenomena arising from this action, and shall here subjoin a few more equally curious. The patients submitted to this kind of magnetisation experienced various unusual sensations, such as undefinable pains in the body, particularly in the head and stomach; an increase or suppression of cutaneous perspiration; palpitation of the heart, and a momentary obstruction of breathing. Sometimes a certain exaltation of the mind, and a lively sense of comfort was experienced; the nervous system in particular was often powerfully affected. The organs of the senses also underwent unusual modifications; ringing in the ears, vertigo, and sometimes somnolency of a peculiar character, supervened. These effects, which were diversified according to the nature of the diseases and idiosyncracies of the patients, became more and more developed as the operation proceeded; and this series of phenomena terminated by convulsions. This convulsive state being established in one patient, soon manifested itself in all the rest. It was designated, as I have already explained, by the name of the magnetic crisis; and as this usually was the *ultimatum* of the results produced, it was looked upon as the chief object of the magnetic influence, and as the means employed by nature to effect a cure. Sometimes these convulsions were of remarkable violence and duration, and on this account patients seized by them were carried into an adjoining room, called *salle des crises*, where they gradually recovered their senses; and then, which is a remarkable circumstance, they experienced no inconvenience, excepting a slight sensation of fatigue; many even stated that they felt decidedly relieved. To these physiological, were often added very extraordinary moral, phenomena; some of the patients burst into immoderate fits of laughter, others melted into tears; they often appeared mutually attracted by irresistible impulses of sympathy, and seemed to entertain the most lively affection for each other. But the most surprising circumstance was the prodigious influence which the magnetiser exercised over his patients. The least sign of his will excited or calmed the convulsions, commanded love or hatred. He thus stood before them like a magician with his wand, under the waving of which their souls and bodies were kept in submissive obedience. Such were the results of the magnetic operations as conducted by Mesmer, which the ancient commissioners verified, and minutely described in their reports.

But as the science progressed, these processes were discovered to be superfluous, and the theory of the universal fluid underwent a considerable modification; inasmuch as the agent was soon believed by the magnetisers to be a peculiar vital fluid, secreted or accumulated more or less in the brain, and of which the nerves acted as the conductors. This fluid was described as being subservient to the will of the magnetiser, who, by his act of volition, could control its influence, propel it externally or internally, and direct it so as to accumulate in any part of the living body.

In accordance with these views, the following are the directions which the Marquis de Puységur gave for conducting the operation: "You are," said he, "to consider yourself as a magnet; your arms, and particularly your hands, being its poles; and, when you touch a patient by laying one of your hands on his back and the other in direct opposition upon his stomach, you are to imagine that the magnetic fluid has a tendency to circulate from one hand to the other through the body of the patient. You may vary this position by placing one hand upon the head and the other upon the stomach, still with the same intention, the same desire of doing good. The circulation from one hand to the other will continue, the head and stomach being the parts of the body where the greatest number of nerves converge, these are, therefore, the two centres to which your action ought to be mostly directed. Friction is quite unnecessary; it is sufficient to touch with great attention, endeavouring to produce an increase of heat in the palms of the hands, &c. All the magnetic effects are more or less beneficial; one of the most satisfactory is somnambulism, but it is not the most frequent, and some patients may be perfectly cured without entering into this particular state. We should not always be intent upon producing somnambulism, for the desire of obtaining any particular result is frequently the cause of no effect whatever being produced. No magnetiser should blindly rely upon nature for the proper regulation and control of the effects which he expects will result from his magnetic operation."

M. de Puységur further adds, "If, when magnetising a patient, you perceive that he experiences a certain numbness, or slight spasm, attended with nervous shocks; and should you then observe that he closes his eyes, you must rub them lightly with your thumbs to prevent the convulsive winking of the eyelids. You will know that your patient is in the magnetic sleep when you see him sensible to your action when you hold your thumb opposite to the plexus. A patient, in his crisis, should answer no one but his magnetiser, and allow nobody else to touch him."

"The somnambulic state requires the greatest caution. Man in the magnetic state is to be considered as the most interesting being in existence. With regard to his magnetiser, it is through his unbounded confidence in you that you have been enabled to bring him completely under your control. It is, therefore, for no other purpose but that of benefiting him, that you have any right to exert your power. Attempting to deceive him, or abuse his confidence, while in this state, is to commit a dishonest action, having a tendency contrary to his benefit; whence it follows that a contrary result is produced to that originally contemplated. The patient must not be annoyed with questions; he must be allowed to take cognizance of his new state. As it is by an act of your will that you put him to sleep, so it is by an act of your will that you can awaken him. It may happen sometimes that the patient will experience trembling, or slightly convulsive motions; in that case, you must immediately stop your first operation, and give all your attention to the alleviation of his sufferings. There are various degrees of somnambulism. Sometimes nothing but a simple somnolency can be obtained from one patient; in another, the effect of magnetism is to close his eyelids so that he cannot open them of his own accord; he then hears everybody, and is not completely in the magnetic state. This intermediate state is very common. These effects are not so beneficial as complete somnambulism, for the magnetiser cannot learn anything from the patient; but they are as favourable to health. Some precautions are to be taken with a patient passing into somnambulism. As soon as you perceive that the patient has closed his eyes, and otherwise manifested some sensibility to the magnetic emanation, you must not at first tease him with questions, and, least of all, endeavour to make him do anything requiring muscular exertion. The state in which he is plunged is new to him: you must let him, as it were, take cognizance of it. The first question should be, 'How do you find yourself?' and then, 'Do you feel any benefit from what I am doing?' You may then express the pleasure you feel by giving him relief. Hence you gradually come to the details of his malady, and the subject of your first questions should never go beyond the state of his health. You ought not to act in opposition to the desire of your somnambulist; you should consult him about the time when he wishes to be magnetised and drawn out of his crisis, likewise respecting the medicaments he requires; and always follow his advice with the utmost punctuality. However remote a somnambulist's prescription may be from the general notions previously adopted in medicine, his own sensation is a safer guide than all the *data* resulting from observation. He is, as it were, the interpreter of nature; he follows the dictates of a lucid instinct; not to obey them in every particular would be missing the aim we have in view,—namely, his restoration to health." M. de

Puységur concludes his instructions by reminding his pupils that "man, being ever actuated by the consideration of his own interest, seldom performs a good action without at the same time serving his own private ends. In other words, it is only by discovering in himself a spiritual principle of a higher order, emanating directly from the Creator of the universe, that he can feel the necessity of gratifying the desires of his soul, who, like her own divine Original, can delight in nothing but that which is in harmony with goodness and truth. This intimate conviction increases to a considerable degree the magnetiser's power."

This new method of magnetising modifies, in a very remarkable manner, the effects produced. The dreadful crises before alluded to almost entirely disappeared, the coughing, hiccup, and immoderate laughter, so frequently brought on by Mesmer's treatment, now seldom occurred; and when they did appear were easily subdued. All magnetisers, however, did not agree in the adoption of the new method. Some maintained that gestures performed without the concurrence of the will to act, and even when accompanied by a contrary volition, were nevertheless perfectly magnetic, and actually produced the usual effects. Others, on the contrary, insisted that the will to act should always accompany the gestures; and that these without the will were absolutely powerless. There were even some who looked upon the gestures as a useless accompaniment to the operation, and saw nothing more in them than a mechanical contrivance, calculated to fix the attention and sustain the will of the magnetiser, the latter being considered as the sole cause of the magnetic phenomena. But notwithstanding this conflict of opinions, the manipulations continued to form a principal part of the operation, and the instructions of M. de Puységur were generally followed. He also observed that the patients who did not fall into crises were sooner cured than those who did; and from his numerous observations, he thought he could establish, in contradiction to the theory of Mesmer, that the convulsions were an unnatural state; that, far from being conducive to the cure of diseases, they had an opposite tendency; and, finally, that it was the magnetiser's duty to calm rather than provoke them.

Thus a new doctrine was promulgated, which, unfortunately for animal magnetism, was as little satisfactory as that of Mesmer; nor was it until long after the publication of his memoirs that M. de Puységur discovered the fallacy of his explanations. The experiment which led him to this discovery I shall here adduce. It gave rise to the doctrines which are at present entertained. "According to my usual practice," says he, "I was magnetising a young man by laying one of my hands on his head, and the other on his stomach. After a quarter of an hour's attention and concentration on my part, and perfect tranquillity on his, he told me that he felt nothing. As he had no complaint, this appeared to me quite natural. I however again pressed him between both my hands, merely to try whether I would be more successful; but he felt no more this second time than he had the first. I was at length about to leave him, when, on slowly removing my hand from his stomach, he fetched a sigh, and complained that I was hurting him. As I did not then touch him, I could not at first believe it; but he hastily took my hand and lowered it, saying that it stopped his breath. I quickly brought myself again into contact with him, expecting that he would now feel a more decided sensation, but it proved quite the contrary; the pressure of my hand had no effect whatever. On removing it to a distance of about one foot from him, he again complained; at two feet distance, he felt a weight on his breast, and desired me to withdraw. I then drew myself back by degrees, and stopped only when he told me his pain was gone and he felt nothing. I was then five paces from him; I magnetised him at that distance by a slow and circular oscillation of my hand; and immediately his head reclined on his shoulder, and somnambulism supervened."

From the time of this experiment, which was performed in 1811, M. de Puységur magnetised at a distance; and, in his last memoir, he asserts that a great number of facts left no doubt on his mind of the superior advantage of this new mode of operating. But, notwithstanding all this, I am persuaded that the doctrines which have hitherto been taught are premature, and that, although they are supported by a great number of facts, yet most of those facts have been but imperfectly observed, and the theories based upon them are of necessity erroneous. The absurd statements and extravagant theories which have been put forth on the subject of animal magnetism are almost incredible; and it is by no means wonderful that many conscientious individuals have, on the very threshold of the inquiry, turned away in disgust. It has even been alleged that, in order to be successful, the magnetiser should possess the cardinal virtues of faith, hope, and charity,—that magnetism can take no effect on any patient in the presence of a sceptic; in fact, no discovery has ever been more exposed by its own advocates to the ridicule and satire of its opponents.

Here I might cite a variety of directions for conducting the magnetic operation by the Count de Lutzelburg, the Chevalier Barbarin, M. de Mont-Ferrier, &c.; but those which are most in accordance with the present state of the science are given by M. Deleuze. "When a patient," he observes, "wishes you to cure him by magnetism, and when his friends and medical attendants are not opposed to it, should you feel disposed to comply with his desire, and be determined to continue the treatment so long as it may be necessary, you must agree with him as to the time of each sitting, and impress on him that his promise must be punctual, and that he must not be satisfied with a few day's trial, but must follow your advice respecting his diet," &c.

"Having thus conferred together, and resolved upon treating the subject seriously, remove from the patient all such persons as might annoy you, and keep none but the necessary witnesses (one only, if possible,) in the room. Then prepare yourself so as to be neither too warm nor too cold, and to enjoy perfect freedom in your gestures; you should also take your precautions not to be interrupted during the sitting. These preliminaries arranged, seat your patient as conveniently as possible, and place yourself opposite to him, on a seat rather more elevated than his, so as to hold his knees between yours, and to touch your feet with his own. Request him to give himself up, to think of nothing, and not to distract his attention by examining the effects he may experience; to be full of hope, and not to be uneasy or alarmed, should the magnetic influence produce in him momentary pains. After having composed yourself, hold his thumbs between your fingers, so that the inside of your thumbs may touch the inside of his, and fix your eyes upon him. You may remain from two to five minutes in this position, or until you feel that your thumbs and his are at the same temperature. This being done, you must withdraw your hands, by moving them outwardly right and left, so that the inward surface be turned outwards, and raise them as high as the head; you must then lay them on both shoulders, and leave them there for about one minute; then bring them down along the arms to the extremity of the fingers, touching slightly all the way. You will repeat this manipulation five or six times, keeping your hands off the body when you raise them. You will then hold your hands above the head for a moment, and draw them down before the face, at a distance of about two inches, as low as the pit of the stomach. Here you will stop again for about two minutes, laying your thumbs on the pit of the stomach, and your fingers under the ribs. You will then slowly come down the body as low as the knees. These manipulations should be repeated during the greater part of the sitting. You will also occasionally come nearer to the patient, so as to lay your hands behind his shoulders, and bring them slowly down the spine, and thence over the hips and along the thighs, down to the knees or to the feet. When you wish to bring the sitting to a close, you must take care to draw the magnetic fluid to the extremities of the hands and feet, by lengthening your line of motion beyond these extremities, each time shaking your fingers. Lastly, you will make before the face, and even before the breast, a few transverse manipulations, at a distance of three or four inches. It is essential to magnetise invariably downward from the head toward the extremities, and never upwards from the extremities towards the head. The downward manipulations are magnetic,—that is, they are accompanied with the intention of magnetising. The movements made upwards are not so. When the magnetiser operates upon the magnetisee, they are said to be *en rapport*—which means a peculiar and acquired disposition, by virtue of which the magnetiser exerts an influence upon the magnetisee; in other words, a communication of vital principle is established between them; and when this has once taken place, the magnetic action is renewed at every subsequent sitting, the instant the operation begins."

It would not be easy to give a better description of the mode of conducting the magnetic operation than this of M. Deleuze; but from his very minute account of the manipulations, it might be supposed that the operation consisted principally in them. This would be a grievous error. It is important to remember that the magnetic power does not consist in mere gestures; another medium is necessary, which the manipulations merely bring into play at the command of the will. This medium may be termed the vital principle, life spiritualised, universal, magnetic, or nervous fluid—it matters not. But most assuredly there is an emanation of a peculiar agent, for out of nothing nothing comes; and manipulations alone, performed with the greatest care, could never produce the slightest magnetic results. The operation may be said to be almost purely intellectual—its success depending on the energy of the will; and we must learn to exercise this faculty before we can educe its latent energies. Hence we observe some magnetisers readily obtain results by processes much less complicated. They merely move their hands in a certain direction, as taught by Dr. Rostan. They excite the course of the nerves by slightly touching the individual alternately on the head and epigastrium. I know many who never touch their patients; they act at a distance of some inches, and have a particular rule of conduct to control the effects produced; and in the cure of diseases this method appears to be more successful than that accompanied by complex manual movements. The error of magnetisers in general is, that they pursue the arbitrary dictates of their own fancy, instead of being guided by observation. They aver, with much self-confidence, that if we wish to produce a certain effect, we must magnetise in such or such a manner; that if we would obtain another result, we must operate differently. If, taking their rules for our guidance, we wish to produce somnambulism, it is necessary to magnetise the patient's brain; but, far from increasing the tendency to sleep, this very manœuvre often prevents it altogether, by producing an over-excitement, which keeps the individual in a state of greater watchfulness than that which preceded the operation;* whereas, had the same individual been magnetised without exciting the head more than any other part of his body, sleep might have been induced. Nay, it is even sometimes necessary to magnetise organs remote from the brain; for they exercise a sympathetic influence on this organ, which brings on the crisis more speedily and certainly than if it were itself magnetised.

The same thing happens with a certain secretion,

* The *will,* said a somnambule, should always *yield* the *command* to nature.

which is supposed to be facilitated or retarded by certain processes, whereas the result often proves the reverse of what was intended. So, also, in endeavouring to awake a somnambule, I remember often having been much perplexed; for it is an established rule among magnetisers, that a somnambulist be awakened by transverse manipulations over his eyes and face. But it has often occurred to me that I have been unable to awaken, and obliged to leave the somnambulist in his sleep, notwithstanding I had employed all the means prescribed in similar cases. In vain have I rubbed the eyelids, even so much as to produce ecchymosis on this very sensible part; still sleep continued much beyond the period I had assigned; and what is still more remarkable, its intensity was greater than when the regular magnetic sleep had not been disturbed.

The wisest magnetisers are those who unassumingly follow the course of nature, and do not presume to control her by laws of their own fabrication; but they are few, indeed, compared with those who strive to outvie each other by feats of skill, and unlimited exaggeration of their own individual power; but this vaunted power has nevertheless its boundaries, which cannot be overstepped, for, were it otherwise, man would at once become a Deity. The practice of magnetism has, however, improved: we now seldom witness, in the treatment of diseases, the exhibition of those dreadful crises formerly considered critical, and deemed necessary. They are now confined to a very few cases, and calmed as soon as they appear, which, generally speaking, is the most proper course.

Every magnetiser ought, as a physician, to know

that their intensity should in some cases be increased, in some diminished, in others provoked, when nature, sinking under the weight of disease, is almost powerless to effect a reaction. It is by observations of this kind that we can alone assist nature in her struggles against the morbid action which is lighted up by disease; but this power of directing the crises can only be practised by one who has made magnetism his exclusive study, which is the case with few magnetisers, many of whom do not know whether they are doing good or evil.

But to return. It is a great mistake to suppose that magnetism only acts on weak and nervous persons. I have often magnetised men of the most robust habit, and produced on them more remarkable effects than on persons of debilitated constitutions. No one should ever be considered an unfit subject; and no person an incompetent observer of the magnetic operations. The unbelief of one individual is no more an impediment to the production of the magnetic effects upon himself than upon a third person. I have often magnetised sceptics themselves, while they were surrounded by none but sceptics; so true it is, that when we take up the defence of truth we must overcome our natural timidity, and follow the maxim of Epictetus, who observes,—"When you have come to the conviction that a thing can be done, do not fear being seen when doing it, notwithstanding everybody should say that you are in the wrong. If you do not act thus, avoid doing it at all; but if you will do it, you should never fear the unreasonable censure of men."

Were it necessary to unite all the qualifications required by some theorists to be a good magnetiser, there would be few such indeed. It is by no means essential even that he should be of a good physical and moral constitution, perfectly healthy, or of a gentle disposition, &c.; for, however, good and desirable may be these qualities, they are not indispensable. I have seen individuals of a very defective constitution, who were even deformed, and of delicate health, obtain, by their magnetic power, the production of phenomena which could not always have been elicited by men of the strongest habit of body,—

"Chacun de cette flamme obtint une étincelle."

Some magnetisers imagine that they possess greater virtue, and magnetic powers more developed, than other men; but this is an absurd notion, which might revive the idea that some are in a manner possessed of some supernatural or occult power similar to that which led to the stake so many wretched beings whose only crime was that of having discovered within themselves a faculty possessed by all, but which had remained latent until certain favourable circumstances accidentally disclosed it to their observation. We need not refer to the remotest antiquity to find numerous instances of these facts; we have merely to open the records of the barbarous times which preceded the revival of letters, and to follow the subsequent historians of more enlightened days, in successive order; and what do we find there? Regular indictments and trials on charges brought before superior courts against pretended sorcerers. Those indictments were based upon well-attested facts; the accused themselves confessed to them; witnesses were brought to corroborate them, who conscientiously stated on oath that which they had seen; and these unfortunate beings were then formally condemned, and suffered the penalty of death in the most excruciating agonies.*

Here I shall perhaps be excused a slight digression to introduce the following anecdote related by Grégoire de Tours, which will doubtless suggest its own moral:—" A wood-cutter of the vicinity of Bourges, had entered a forest for the purpose of cutting wood, when he was suddenly surrounded by a swarm of bees, and literally covered with their stings, which left him mad, or in a state bordering upon madness, in which, for the next two years, he continued. After this adventure, the woodcutter, having passed through the neighbouring towns and villages, went to the province of Arles. Here he covered his body with skins, and lived as an hermit, praying night and day; but in order to deceive him, Satan transmitted to him the power of divining. He soon left the province of Arles, and proceeded through the Gevaudan. Here the people crowded round him, bringing to him the sick and infirm, whom he cured by touch. Gold, silver, and garments, were given to him; but he distributed the whole among the poor, prostrating himself on the ground, and praying incessantly. He foretold the future, announcing to some; diseases, to others, losses. All this he did," says Grégoire de Tours, "through the Satanic art, and some, unnatural prestiges. He seduced a great number of people, not only country people, but even clergymen themselves."

A few baths and bloodlettings might have cured

* Father Sprée, a Jesuit, did not himself hesitate to protest against the manner in which sorcerers were prosecuted. He affirms, that out of many sorcerers whom he exhorted and led to death, there were some of whose innocence he was as perfectly convinced as of his own existence.

this man, supposed to be possessed by the devil. Aurélius, bishop of Puy, however, contrived that he should be surprised, and put to death.

But to proceed. It is not, as I have just stated, necessary that an individual should believe in animal magnetism, to experience the effects which it induces. If the patient be perfectly passive, he is generally more susceptible of being magnetised with success than in any other condition. Furthermore, contrary to that which has been often stated, the passes serving to conduct the magnetic fluid may be made either perpendicularly or transversely. The sensible or apparent effect, however, seems to be produced only when there is an incipient or complete saturation of the nervous system of the patient, by the fluid which apparently issues from the nerves of the operator. The most favourable condition to receive this magnetic saturation with benefit, is not yet accurately determined; a weakly or abnormal state is, however, generally speaking, most favourable to its reception. On the part of the magnetiser, the most important rule he can adopt is to exert the greatest energy of volition he can command. As, when the light of the sun is transmitted through a burning glass, even in the depth of winter, the solar rays which previously gave little or no warmth, being concentrated and thus brought to a focus, ignite the combustible substances exposed to them: so likewise the human mind, which is the mirror of the soul, by converging its rays into one focus, affects the soul brought into juxta-position with itself. I repeat, the magnetiser must *will* with the utmost perseverance; he must not pity when he can succour the afflicted; he need offer no vows; but let him believe in his power, and act with energy. I do not mean with violent mental excitement, for this will neutralise the effects, by absorbing the principle which ought to produce them. He should, on the contrary, enjoy perfect ease and freedom ; and though he is to send to his extremities a momentum or force sufficient to raise a considerable weight, he must have nothing but his own limbs to raise. It is the excess of this momentum which strikes the patient, and produces all the magnetic phenomena. I would even advise him not to divide his attention by thinking of the manipulations prescribed by M. Deleuze, for by them alone the phenomena cannot be induced ; besides which, when he looks for instruction, he will be at a loss to determine whether the effects produced be the result of the monotony of the process. I also recommend the magnetiser to stand before his patient instead of sitting. He should not touch him, but his passes should be made opposite to him, at a little distance. Every magnetiser does not, however, obtain the same results. These impediments in general may arise from a natural want of power in the magnetiser ; and this may be induced by a lesion or debility of his organs ; or it may proceed from his mind being unable to act freely, either through the same vitiated disposition of his organs, or because his fancy is too quick, and easily passes from one subject to another.

A greater obstacle to the development of the magnetic power is, pride. One of my somnambulists once observed that a man may possess an independent fortune, which may allow of his time being devoted to the relief of the poor, to which he may add all the conditions of disinterestedness, persevering application, strength, charity, &c. ; and yet he may not be entirely successful, because in his heart the pride of success unnerves his faculty of doing good. A variety of incidental circumstances, which it is even difficult to enumerate, an over-anxiety to produce the effects, or any incidental suggestion that may disturb the attention of the magnetiser, will often be sufficient to mar the successful issue of the experiment; and to some such cause, I doubt not, may be attributed the recent failure of M. Berna in endeavouring to produce some of the higher magnetic phenomena before the French Academy. But we need not a formal report of the Academy to prove that the operation of animal magnetism is often unsuccessful; every tyro in logic knows that such negative evidence as this is utterly worthless. In public lecture-rooms the professors of the physical sciences constantly fail in a variety of delicate experiments; but is it ever argued that such failures invalidate the principles of the science which they intended to illustrate? Certainly not. One positive fact is not to be repudiated by a thousand negative attempts to reproduce it. Such an argument as this cannot for a moment be entertained. He, therefore, who may fail in attempting to produce the magnetical phenomena, must not on that account deny their existence; let him return to the trial, and I have no hesitation in assuring him, that if the preliminary conditions required by the magnetisers be complied with, he will eventually succeed.*

* Vide Report on Animal Magnetism made to the Royal Academy of Paris. London Medical Gazette, vol. xxii. p. 913 *et seq.*

CHAPTER V.

PRECAUTIONS TO BE OBSERVED, AND DANGERS TO BE APPREHENDED, DURING THE MAGNETIC OPERATION.

Having described the most eligible method of conducting the operation of animal magnetism, I am anxious to impress on the minds of those who may feel inclined to try the experiment, that the operation is not always unattended with danger. Hitherto I have appealed in a very earnest manner to the integrity of the evidence in favour of the science; but so immediate are the proofs which every individual can command, that I feel it incumbent on me to caution sceptics themselves against trifling with so powerful an agency; for I have known instances of many who, in endeavouring to induce the magnetic phenomena, have placed themselves in a very painful position, and the person operated on in a very alarming state. Of course animal magnetism, like every other science, has its own laws, and these should be diligently studied before any individual attempt to practise it. The constitution of the patient, whether in health or disease, should also be well understood; indeed, physicians alone should be entrusted with its application.

When a person magnetises for the first time, he is generally actuated by a feeling of curiosity, blended with doubt and apprehension. If the patient he magnetise, under the influence of these feelings, happen to be sensible to his action, he will soon manifest symptoms which will greatly surprise him; and if the magnetiser possess no method of controlling them, they may assume a most alarming character. The patient, perhaps, will fall into a state of most violent convulsions, and the more the ignorant magnetiser endeavours to relieve them, the more will they increase in their intensity and apparent torture. Such convulsions have been known to last from six to eight hours without any intermission; and the patient thus affected has continued seriously ill for many successive days afterwards, feeling a sensation as if his limbs were broken, accompanied by so profound an aversion for magnetism and the magnetiser, that the mere mention of either superinduces a return of the fits.

M. de S. C., a retired officer, having heard a vague report of animal magnetism, attempted to make the experiment upon his own daughter, although she complained of no illness. He merely wished to ascertain whether he could make her feel the magnetic sensations. With this view, and without being aware of the extent of mischief he was provoking, he laid his hand on the stomach of his daughter, and obeyed the magnetic injunctions. After a few moments of magnetisation, she experienced spasmodic attacks, which, far from alarming her father, only encouraged him to proceed. In a very short time, however, Mademoiselle de C. was seized with the most violent convulsions; and her father now, not knowing the proper way to calm them, only increased their intensity. In much distress, he was obliged to leave his daughter in this state, and she spent the following night in incessant convulsions, and remained a whole week in this dangerous state.

Another case of this kind was transmitted to M. de Puységur by M. Segrettier, landowner at Nantes. The following are the particulars :—" A young lady of distinguished birth, who seemed to enjoy most excellent health, happening to be on a visit at the chateau of her relative, the Marquis de B., was indulging with the rest of the company in passing sundry jokes upon magnetism. Her uncle, M. de B., who outstepped by his sarcastic remarks everybody present, and was gesticulating with great freedom, began to direct his pretended influence upon his niece, when they both set about magnetising each other as fast as they could. At first the young lady laughed very heartily, but it was soon discovered that this laughter was anything but natural; and the first surprise excited by this phenomenon soon gave way to unspeakable terror, when it was manifest that she was gradually losing her reason and the use of her senses. Indeed, she could no longer see, nor hear, nor speak ; her eyes were immoveably fixed; her neck, outstretched, resembled a weaker magnet violently attracted by a stronger one ; she followed her magnetiser everywhere, and yielded to his sole influence. The spectators attempted to separate them, but this only provoked dreadful convulsions. Her magnetiser, on his part, felt extraordinary sensations, which, in addition to the shock he had experienced by the alarming state of his niece, had entirely altered his features, which became extremely pale and dejected. In the course of a few hours, the crisis of the magnetised young lady gradually ceased, and she complained of acute pains in her stomach. The remainder of the day, and the following night, were passed alternately in convulsions and magnetic sleep ; and this state lasted several days."

The magnetic operation, under such circumstances, produces, I repeat, the most alarming results; the individual subjected to the experiment feels, by degrees, a torpor creeping over his limbs ; he loses the consciousness of their position ; and if the magnetiser persist in his action upon him, the thoracic muscles may be seized with a temporary paralysis, followed by an impeded action of the inspiratory muscles, and a kind of mucous rattle becomes audible. In vain the patient implores assistance ; and if the operator know not how to induce a cessation of this state, the patient may incur the greatest danger. I have seen several cases of this kind, and I know that in one of the chief hospitals in Paris a similar instance occurred, and the physicians present, not being aware of the existence of so singular a phenomenon, were exceedingly alarmed. They found, on attempting to release the patient from her oppression, that it was out of their power, and for about thirty minutes, at least, they could obtain no cessation of the paralysis. In that short period the paroxysm had run through every stage of asphyxia. Respiration at first became intermittent, and then imperceptible ; the skin was discoloured, the veins unusually swelled ; there had been loss of memory during several minutes. This crisis happily terminated favourably ; but all those who witnessed it were of opinion that, if the obstacles by which the circulation had been impeded, had resisted a little longer, the patient must inevitably have died.

In the usual magnetic sleep, there sometimes occurs a peculiar state, seldom observed, and much dreaded by magnetisers, because very few among them possess the means of bringing it to a favourable issue. I will endeavour to describe it. A person plunged in the magnetic sleep (and it generally occurs in those individuals whose sleep is most profound) collapses into an extraordinary state, of which the following are the chief symptoms:—The somnambulist who heard his magnetiser perfectly, is suddenly seized with a fit of deafness; he neither hears nor feels him any longer; and the magnetiser discovers that he has entirely lost all control over him; he no longer obeys his injunctions; he is as dumb to him as to every body else; his jaws are firmly fixed together, and it would be easier to break than sunder them; he is motionless, yields to the laws of gravitation, and his body appears powerfully attracted to the ground. The pulse at the wrist diminishes its beats in number and intensity; the temperature of the body perceptibly lowers, and inevitable death appears to be impending. If the magnetiser be familiar with this phenomenon, and do not abandon the patient, he will gradually recover from this state of concentration; the pulse will resume its usual cadence, and, returning into the ordinary state of somnambulism, he will, on being questioned, state that he has had certain visions during his lethargic state; but, by a strange anomaly, although still in the somnambulic state, he can hardly recal them, or even describe the sensations he experienced.

No symptom can indicate the moment when this crisis supervenes; I have observed it a great many times. It has often occurred when I intended to bring on a cessation of the usual magnetic sleep, that the somnambulist suddenly, and against my will, collapsed into this singular condition, in which he remained for several hours together. It is the most extraordinary state known in animal magnetism; it is, perhaps, the most dangerous, and also the best calculated to afford instruction, when we know how to question, in proper time, these extatic somnambules. If we could hit upon the ideas which are uppermost in their minds, it is probable that we might obtain the most useful revelations; there is but one moment for this; we must seize the instant of transition from the ecstatic to the usual somnambulic state, for they soon lose all recollection of the sensations they experienced. This state seems to be the limit of an entirely new order of being; "it is an unfathomable mystery, in which the mind is wholly lost;" and I consider it the most dangerous crisis that can occur, and the most likely to be attended by fatal consequences, should the magnetiser leave the patient to himself, after having plunged him into an ecstatic sleep.

"One day," says M. Chardel, "while magnetising a somnambulist, and being full of self-confidence, I allowed her to walk about the room with a friend of hers; they were conversing together, and my attention was then diverted from the extraordinary mode of existence which I had just produced; when the two friends suddenly requested me to rehearse a scene from Racine's Tragedies. I imprudently entered too fervently into the spirit and sentiments which the poet expresses with so much truth, and did not perceive the emotion of my somnambulist until she fell motionless on the floor. Never was loss of the senses more complete; the inanimate body lay stretched on the floor with all the flaccidity of death; each limb, on being lifted up, fell back with all its weight; breathing was stopped, the pulse and beatings of the heart were no longer felt, the lips and gums were discoloured, and the skin, having lost the stimulus of circulation, assumed a livid and yellowish hue. Every thing seemed to indicate that I had nothing but a corpse before me. I happily kept my presence of mind, for the purity of my intention increased the calm but resolute energy of my self-devotion, and I had sufficient control over myself to feel that I could still exert a great power over my somnambulist. I began by magnetising the plexus; I then breathed into her mouth, nostrils, and ears; and by degrees she recovered the use of speech; this speech was at first weak, but it was soon distinct, and in answer to my questions I learned that nothing had injured the health of my patient," &c.

Those who experience these singular effects, far from complaining of them, desire that their duration may be prolonged.

" Why do you call me back to life?" said a somnambulist in her magnetic exaltation: " if you would only go away, this body which oppresses me would grow cold, and my soul would no longer be here on your return. I should then be perfectly happy."

All hold nearly the same language, and suggest the idea of the soul being partially disencumbered of the coils of its mortality—seeing, hearing, feeling, taking cognizance of all things past, present, and future, through some other channels than the physical organs of its subordinate manifestations. All, too, agree in enjoying, in this state, a sort of exquisite elysium of repose from which they dread to be disturbed; the soul, apparently half set free, shrinks from being again brought back and entangled in the chains which bind it down to the narrow sphere of suffering humanity. It is impossible to contemplate a somnambulist in this state without a feeling of awe, not unmixed with anxious wonder; he is a being who appears to belong more to the world that is to come, than to that in which, as finite beings, we exist; he already appears half disrobed of his carnal nature, and participating in the sense, if not the actual enjoyment, of his immortality: it is impossible to divine what views of infinity may now open before him; all we observe is, a being like ourselves exalted into a state of temporary beatifaction, far above our sympathy, and beyond our comprehension. They invariably beseech us not to awaken them or bring them back to their ordinary condition; but the magnetiser should be cautious how he listens to their entreaties, and should, rather than otherwise, hasten to draw them out of this state; for he should remember that he is exhausting his own energies, and the moment may arrive when it might be necessary for further exertion, and he be unable to command it. He should regard himself as being the mere organ, as it were, secreting or throwing out, under the energy of his volition, the magnetic fluid; and in proportion as he becomes fatigued, and the secretion itself exhausted, his influence over the patient, whether in the magnetic state or not, must diminish or altogether cease. On this account it is that magnetisers often fail to induce the effects they anticipate; they exert themselves to the utmost; the patient exhibits all the premonitory signs of the magnetic state, but a step is yet wanting—the step itself is not established —because the operator cannot command the requisite magnetic power. It is, indeed, wonderful how few men can command any high degree of *sustained* mental energy; the faculties of the mind, like the limbs of the human body, require exercise in order that they may attain their full development; but so much are they neglected that the finest susceptibilities of our nature are often not called into action, and may be literally said to go down with us into the grave unborn.

Inexperienced magnetisers are not aware of the difficulties they have to encounter, and know not the dangers which are to be apprehended; they are truly working in the dark, with a power which sometimes, to their astonishment, attains an ascendancy over them, and produces effects beyond their participation or control. Hence many, in trying their own magnetic powers as a matter of sheer curiosity, or *en badinage*, have found themselves suddenly placed in a most perplexing and distressing predicament. The voluntary patient has gradually succumbed under their influence until a state of collapse, simulating death itself, has supervened. In the thesis of M. Fillassier, which was recently quoted in a clinical lecture of M. Andral, a case of this description is related. It will be found reported in the Lancet. "In this inaugural thesis," says M. Andral, "which was defended before our faculty of medicine, there is the narrative of an occurrence bearing irresistible proof of authenticity. The facts are particularly curious. The author, then an *interne* at the Hôtel-Dieu, and totally sceptical regarding the powers ascribed to this mysterious essence, this asserted magnetic fluid, formed, for amusement, the plan, with a brother *interne* equally incredulous, of submitting this friend to the manœuvres of the magnetisers. The *passes* were continued for about twenty minutes without any remarkable effect, but at the expiration of that time the young man began to yawn, his eyelids grew heavy, and closed involuntarily; he attempted to shake off the torpor in vain; his respiration next became accelerated, his head fell on his shoulders, and he uttered a sardonic laugh of indescribable expression. We thought," continues M. Fillassier, "that he was amusing himself at our expense; but, in a little time, what was my horror when I saw his fingers turn blue, his head fall powerless forward, —when I heard his respiration rattling like a dying man's, and felt his skin as cold as death itself! I cannot find words to describe my sufferings. I knew not what to do. Meanwhile, all these horrid phenomena increased in intensity. I tremble at the recollection of what I saw: there lay my friend, my victim, devoid of the aspect of life, in a state of complete and terrible collapse. With his hand clasped in mine, in a state of agony no tongue could tell of, I laid him on a bed, and waited the result in a state of mind I never can forget. In a quarter of an hour he recovered, and, exclaiming that in the ecstasy he had experienced sensations of extreme delight, begged me to recommence the passes. I did so with less apprehension, and again the somnolency proceeded. The collapse, however, was less profound and terrific, and in some minutes he suddenly awoke with the exclamation, 'What happiness is this!'"

As I have already observed, it is sometimes very difficult to awaken or bring patients out of the magnetic sleep; speaking generally, it is more difficult to conduct the operation *after* sleep is induced than it is to produce sleep itself. In most treatises on animal magnetism, it is stated that magnetisers can at pleasure restore their patients to the ordinary, or waking state. This is an error. I have often, in a few minutes, brought on sleep, and could not for hours afterwards awaken the patient, notwithstanding that I have energetically applied all the processes usually prescribed in such cases. In vain have I exerted my abilities to the utmost; the more I have wished and willed to induce the waking state, the more has the intensity of the sleep increased. I have passed my fingers over the eyelids until I have produced ecchymosis, yet all my efforts have proved unavailing; and it is a curious fact, which deserves attention, that when such persons awake from this indomitable sleep they fall again into somnambulism, and then the slightest noise will awaken them. A case of this description was lately communicated to the Lancet by Dr. Sigmond, the Lecturer on Materia Medica and Therapeutics at the Windmill-street School of Medicine in this city. Unlike many of his professional brethren, this gentleman has not disdained to make animal magnetism the subject of conscientious investigation, and has performed himself, to confirm his conviction, several experiments.

"The most remarkable case," says he, "that has fallen under my observation, and which, while it excited in me great anxiety, and the deepest interest, has taught me to prosecute my researches with extreme caution, has occurred to me within the last two days. I was enjoying the hospitality of a most amiable family in Fitzroy-square, when animal magnetism became the topic of conversation, and I related the trials I had already made. One of the young ladies proposed to become the subject of experiments, to which I very willingly assented; for, having on former occasions attended her during momentary sickness, I was fully aware of the natural strength of her constitution, and the absence of that nervous temperament which renders this system totally inapplicable. I began what are technically called 'the passes.' They, as is not unusual, excited laughter and incredulity. I proceeded for about five minutes, and then stopped and inquired if any sensation was produced, and the answer was, 'a slight sleepiness;' and ridicule was again thrown upon the subject. I recommenced the manipulations; I observed the eyelids falling, and at last they closed; but, as the same incredulous smile remained, I persevered for three or four minutes, when I, almost doubting whether any influence had been produced, inquired what the feelings were; to this no answer was returned. I found my young friend was in the most complete trance I had ever yet witnessed as the result of my magnetism. The stupor was most profound; and I then tried the usual means to arouse her, but they were vainly exercised. After a few minutes I found the hands become icy cold, the face lost its natural hue, and became perfectly pallid; the extremities became quite cold; the respiration was imperceptible; the stimulus of light did not affect the eye; on speaking to her, a faint smile was excited, and a quivering of the lower jaw, which seemed to indicate a wish but an incapability of answering; the pulse became gradually feebler, whilst the external appearance altogether bore such a decidedly deathly cast, that naturally some apprehension was excited amongst her family, by whom she was surrounded. Of course I could not but feel a certain degree of anxiety and regret that I had produced such a state, and much uneasiness at the thought that I had inflicted a moment's alarm to my kind friends. These feelings were, however, less acute, from the full knowledge I entertained that the family had long reposed the most perfect confidence in me, and that no member of it had that nervous susceptibility which would have embarrassed me had any untoward accident presented itself.

"I placed the perfectly unconscious subject of this distressing scene in an horizontal position, and directed the application of warmth and of friction to the extremities. Circulation and animal heat were gradually excited, but she presented a most singular appearance of suspended animation. In this condition she remained more than four hours, for I had commenced a little after ten in the evening, and it was about half-past two that, on some slight effort being made to rouse her, she uttered some of the most piercing shrieks I have ever heard; there were convulsive efforts to raise the limbs; the face, too, became convulsed; she opened her eyes, and stared wildly around; she was placed in the upright posture, and seemed sensible. Advantage was taken of this circumstance to carry her to her apartment; before, however, she could reach it, she fell into a profound slumber, but its character was more natural. She was placed in her bed, appearing perfectly composed; the countenance had acquired its natural hue; the respiration was perfectly easy, and the pulse natural. In this state she remained during the whole of the day, until nine o'clock in the evening, once only opening her eyes, and addressing a few words to an anxious and affectionate sister, who never left her side. In the evening, the young lady joined her family, perfectly restored to her wonted cheerfulness. She expressed no complaint whatever. She stated that the feelings that first came over her were those of extreme quiet and repose,—a species of ecstasy,—a gradual languor seemed to steal over her; that she heard something passing around her; felt an inclination, but an utter impossibility, to reply. The first waking up she, however, described as almost terrific. It was as if she was bursting from a narrow and confined space, and as if she arose from interminable darkness. The lesson that I have thus learnt will not be lost upon me."*

This fact deserves attention, inasmuch as it elucidates very clearly the impropriety, and even danger, of sporting with a power which may produce such alarming, and, for the period, uncontrollable symptoms. In vain were physical *stimuli* here applied; a purely psychical series of effects can only be psychically dealt with; and what avail frictions applied to the surface of a body sunk into a state of magnetic insensibility? Have we not seen the skin and flesh torn, lacerated, burned, without exciting the slightest consciousness or impression on the nervous system? Assuredly; and magnetism itself is the only therapeutical agent that could affect a patient in this condition.

Here, also, it should be observed, that, so immediate and powerful is the magnetic influence, that individuals placed near those who are being magnetised sometimes experience all the effects of direct magnetisation. I have often produced somnambulism in this manner. Not long ago, having been called upon to attend a child, the son of Count de Bastard, I was at once informed that the patient was very ill, and that my assistance had been sought as a *dernier ressort*—as the only and last resource which promised them a ray of hope. A young woman, who was much attached to the infant boy, held him in her lap; and in this posi-

* The Lancet, vol. i. p. 388, et seq., December 1837-38.

tion I proceeded to magnetise him, surrounded by a few anxious relatives, who would have given all they possessed on earth for him to be restored. A feeble lamp lighted this melancholy scene, and a profound silence was observed. My influence directed on the patient, at a distance of about one foot, soon produced a salutary effect. The pains which the child had experienced, and expressed by cries and convulsive motions, were soon calmed, and he became comparatively tranquil. It was, however, otherwise with the young person who held him in her lap. Her eyes having closed, she fell into a state of complete somnambulism. Large tears trickled down her cheeks, and she gradually passed into the higher degree of lucidity. Then, when we attempted to take away the child from her, she refused to part with him, and only, after some altercation, gave him up at my request.

The magnetic influence, when thus communicated, instead of inducing sleep, generally gives rise to spasms, convulsions, and alarming symptoms. On another occasion, I was magnetising a lady who was but little sensible of the magnetic influence, when a person who stood beside her felt, during the operation, her limbs benumbed, and pricking pains in her eyelids. On the following day the same lady was again magnetised, without the intention of producing an effect on any one else; but the same attendant again experienced symptoms more decided than on the preceding day, and remained the whole of that day afterwards in a state of excessive lassitude. Again, at the next *séance* she was affected so much that somnambulism was brought on, and yet I had no intention of producing this result. She soon awoke; but, yielding to sleep again, she remained in this state for several hours. A similar incident occurred to me while operating at the house of a patient of mine, the wife of a counsellor of state. Her lady's-maid, who attended the sitting to render assistance if necessary, was seized, while I was magnetising her mistress, with severe spasms, which soon became very alarming. I rose to subdue them, but she hurried away, declaring that I gave her pain, and that she would not allow me to come near her. She, however, soon recovered, and I attributed the effects only to an excited imagination. When perfectly recomposed, she sat down again near to her mistress, upon whom I recommenced the magnetic operation as before. But in about five minutes, and without any intention on my part to exert any influence on this attendant, she was again seized with the same description of paroxysm, accompanied with considerable pain and difficulty of breathing, from which she did not recover for several days afterwards.

It is this indirect mode of magnetisation which doubtless gives rise to anomalies in the effects produced, which would not appear were the same individual subjected to direct magnetisation. The subtlety and power of the magnetic fluid is almost inconceivable. As I have before stated, I believe that no one ever submits to the operation of magnetism without some change being induced in his organisation. He may not be sensible of it at the time; he may rise from the *séance* apparently unaffected; yet some hours afterwards may the magnetic effects develop themselves. This has often occurred to me. I have magnetised persons who were not conscious either during or immediately after the operation, of any results, yet have a short time afterwards become somnambulic. The subtle and penetrating intensity of the magnetic fluid, annihilating all relations of space and time, is, however, more distinctly manifested in the magnetisation of persons at a distance, or through intervening partitions—as through screens, folding doors, or thick walls. The success, however, of this experiment will depend, not only on the power of the magnetiser, but also on the peculiar susceptibility of the patient, which should always be taken into consideration. As a general principle, the stronger, speaking magnetically, magnetises the weaker,—women are more easily affected than men—persons who are sickly than persons in robust health.

It was observed by Deleuze, that the difference of the sex has little or no influence on the magnetic power; that women magnetise as efficiently as men. Their action is, however, more gentle, albeit, not less salutary; and when they possess the requisite confidence, they can, he assures us, magnetise their own sick children with more success than could be done by the most experienced magnetiser. He relates the case of a lady who was afflicted with a very serious complaint, and, despairing of relief from the ordinary resources of medicine, she applied to a magnetiser, who agreed to undertake her treatment. He accordingly did magnetise her during three or four months, and, when visiting her one day, said he was afraid that he should not be able to magnetise her, as he was himself very much indisposed. She then proposed to magnetise *him*—he consented—and in a quarter of an hour he became somnambulic. He requested that his eyes might be bandaged, as he thought he should then be able to see better. The lady then requested that he would examine himself, and endeavour to discover how his sufferings might be relieved. He answered, "I have too little fortitude, and am easily affected. My complaint is trifling, and I shall be well to-morrow. It is you that I must examine. But how fortunate it is that you have made me somnambulic! Henceforth, we shall always commence our magnetic treatment in this manner, and I assure you that you will be well attended to." From that period the lady constantly set her magnetiser asleep every day, and in consequence of following his prescriptions, while in a state of somnambulism, her complaint (a *scirrhus*) was removed. While in this state, too, he predicted that she should be attacked, some time afterwards, by a complaint of a different nature. This prediction was fulfilled, and she was again cured in precisely the same manner as before. The veracity of this account is vouched by M. Deleuze himself, who knew the parties.*

It is laid down as a rule by Deleuze, that women should magnetise women; but I am of opinion, without throwing any doubt on their ability to do so, that

* In the short but interesting memoir given by Captain Medwin, of Percy Bysshe Shelley, we find that he was on one occasion magnetised by his biographer, and on another by a lady, to whom the poet afterwards addressed some very beautiful lines, supposed to have been spoken to him by the fair operator. They begin,—

"Sleep on! sleep on! forget thy pain,
My hand is on thy brow,
My spirit on thy brain,
My pity on thy heart, poor friend,
And from my fingers flow," &c.

These lines may be "hastily thrown together," but breathe all that exquisite sensibility and tenderness which so painfully characterise every fragment he has left behind him. It may be added, that after being magnetised by Captain Medwin, Shelley became an habitual somnambulist.

they should not magnetise patients susceptible of passing into the higher degrees of magnetic extase, because of the dangerous symptoms which may, as I have above explained, supervene, in which case women are certainly more liable to lose their self-command than men.

It is not easy, as I have already premised, to describe the physiognomy of a powerful magnetiser. I have seen men debilitated by disease, emaciated, labouring almost even in the last stage of pulmonary consumption, most powerful magnetisers; it is a mental rather than a physical exertion, which demands great concentration of thought and sustained volition. If his thoughts wander during the operation; if his attention be divided, and he become absent; if even, from over anxiety to induce magnetic results, the equable balance, or steady power with which he should act be disturbed, the effects are more slowly induced, or perhaps fail to appear. Self-confidence in his magnetic power, on the part of the magnetiser is important, if not essential to the success of the operation; and a lively faith in his influence on the part of the patient facilitates very much, and accelerates the development of the effects. This is by no means surprising. In the most common medical cases it is notorious that if the patient possess no confidence in the physician, his prescriptions will rarely prove availing, so direct and powerful an influence has the mind itself over the action of the most eligible therapeutical agents. Hence it is well observed by Burton, in his curiously learned work, the Anatomie of Melancholy,* that "a third thing to bee required

* The Anatomie of Melancholy, fol. Oxford, 1638.

in a patient is confidence, to bee of good cheare, and have sure hope that his physician can helpe him. Damascen, the Arabian, requires likewise in the physician himselfe that he bee confident he can cure him, otherwise his physicke will not be effectuall, and promise withall that he will certainly helpe him—make him beleeve so at least. Galeotus gives this reason, because the forme of health is contained in the physician's minde; and as Galen holds, confidence and hope doe more good than physick; he cures most in whom most are confident. Axiocus, sicke almost to death, at the very sight of Socrates recovered his former health. Paracelsus assigns it for an only cause why Hippocrates was so fortunate in his cures, not for any extraordinarie skill he had, but because the common people had a most strong conceipt of his worth." So also Avicenna affirmed, that "he preferred confidence before art, prescriptions, and all precepts whatever."

There are many persons naturally very susceptible of the magnetic influence. They cannot endure the operation when steadily and vigorously conducted; the accumulation of the magnetic fluid in their system produces an alarming disturbance of all the organic functions; the action of the heart becomes increased, the breathing hurried, violent convulsive fits supervene; and should such persons be labouring under any aneurismal affection, disposition to apoplexy, or far advanced phthisical diseases, the consequences may be fatal. As a general rule, the susceptibility of the patient increases with every successive sitting. He who at first resisted the operation for a considerable time, at length speedily, even in a few seconds, falls into somnambulism; and when the magnetiser has invested him with an extreme aptitude to pass from one state to another, the least gesture from him—a look indicative of his intention—his will expressed with energy, is sufficient. It should also be observed, that during natural sleep persons are magnetised with as much, or perhaps more, facility than when awake. Under the magnetic influence they soon become restless and agitated, and the natural passes into the magnetic sleep, characterised by all the phenomena which are so peculiar to this state of being.

When such convulsions as those above described result from the magnetic operation, the magnetiser should maintain his self-possession; he should not, however much the patient may appear in a state beyond his control, suspend the process of magnetisation, but persevere steadily until he succeed in restoring him to a calmer magnetic state. In some cases very great cerebral excitement, amounting almost to temporary mania, is produced; and he should then continue operating with the utmost mental energy: but it is difficult, if not impossible, to anticipate all the abnormal manifestations which may appear during the process of magnetisation; they require to be watched with vigilance, and at once treated with decision. He, however, who would be a successful operator, should possess medical knowledge; he should know the physiological and pathological condition of the human body, and be prepared to detect the slightest aberrations in the functions of organic life. He will otherwise be incompetent to observe the effects of magnetisation, which should, in a great measure, practically guide him in the mode of conducting the operation, a knowledge which is not to be acquired without much experience, yet which is very essential to the safety of the patient. Hence those continental governments which prohibited the practice of animal magnetism by all, excepting persons duly qualified, acted wisely; for there is no therapeutical agent more liable to be abused on account of its being under the command of every individual; so that the unlearned as well as the learned, sceptics as well as believers, for the purpose of satisfying curiosity, or amusing a group of idle spectators, have frequently produced the most alarming symptoms. In the hands, however, of well-informed and skilful magnetic practitioners, such dangers are not to be apprehended; and this is no more an argument against the application of magnetism, when competently conducted, than the abuse of the lancet would be an argument against the use of this important little surgical instrument. It has been said that the injudicious use of the lancet has produced as great a mortality among the sick, as the sword has in the field of battle; and every medical practitioner knows what fatal consequences have ensued from the ignorant and empirical administration of opium, mercury, antimony, arsenic, preparations of lead, &c.; but these valuable remedies are not, for this reason, to be excluded from the pharmacopœia. So is it with animal magnetism. From the facts above stated it is evident that it is a very powerful agent, which should not be had recourse to excepting by persons who have studied, and have a practical knowledge of the subject. Here, therefore, let me caution sceptics themselves to beware how, in the midst of their incredulity, they sport with an agent that may, when they least expect it, give rise to consequences which they themselves will not contemplate without poignant sorrow and alarm. The power of magnetisation is common to us all, but the method in which it is to be efficiently developed, so as to produce salutary results, demands much study and patient investigation.

CHAPTER VI.

CURATIVE EFFECTS OF ANIMAL MAGNETISM.

If such be the dangers which are to be apprehended from the misapplication and abuse of animal magnetism, what are the advantages to be derived from this discovery? This question is easily answered. The most dangerous remedies when misapplied, are often the most efficient when judiciously administered;—from the most poisonous plants may virtues be extracted. The earth is not to be regarded as altogether an howling wilderness of disease and crime, the paths of which are beset only by noxious reptiles and deadly weeds. A more cheerful view opens before the contemplation of the philanthropist, who will perceive that the current of all things flowing onwards is redolent of blessings. The value of such a discovery as animal magnetism is to be estimated, not by the evils to which its unskilful application may give rise, but by the positive good which may be derived from it. Already we have seen that during the state of magnetic insensibility the most painful surgical operations may be performed, and the patient remain the whole time in a state of perfect unconsciousness. Is this not a boon to suffering humanity? But this is not all; the most obstinate and painful chronic diseases, as well as numerous spasmodic nervous affections, have been relieved and perfectly cured by its application. It was indeed the cures which were wrought by the early magnetisers, Mesmer, Puységur, Deslon, Bickers, Olbers, Wienholt, &c., which first attracted the attention of men of science to the discovery: the attestation of these by competent observers, and the patients themselves, put magnetism in the first instance on so secure a footing as to make head against the sea of prejudice by which on every side it was beset. It was, I repeat, the successful treatment and cure of diseases, which had notoriously resisted every other remedy, which compelled the rudest and most inveterate of our antagonists to recognise the influence of magnetism; and when these facts were demonstrated beyond all reasonable controversy, it remained for them to seek in the umbrage of their imagination the solution of the mystery. Hence imagination, imitation, manual contact, and numerous other hypotheses, equally unsatisfactory, were conjured up; but no rational mind could fail to observe that the appeal was made, not to theory, but to fact; inasmuch as men of high intellectual attainment and unimpeachable moral integrity, who had no interest whatever in the promulgation of animal magnetism, after subjecting themselves to its treatment, declared themselves relieved and cured of various distressing affections. Hereafter I shall prove that medical science is founded on animal magnetism; that the whole art and practice of medicine in the ancient temples consisted in the magnetic treatment which is even at the present day practised in India.* Indeed, no manner of doubt

* "Several of the Indians," says Hearne, who appears to be not at all aware of the *principle* upon which the fact depends, "being very ill, the conjurors, who are always the doctors, and

can exist, that even in the hands of unskilful practitioners, in the midst of all their strange and extravagant superstitions, such men have wrought many very curious and signal cures. It is idle to suppose that any nation would believe generally in the efficacy of any therapeutical agent, were it only after all an errant folly or fiction; for however uninstructed men may be, they derive knowledge from experience, and it is not reasonable to presume that any method of treatment would for ages be persevered in, if the sick therefrom derived no description of benefit. But apart from all argument deduced from the early history of animal magnetism, we now find that in many of the great public hospitals, at Berlin, Stockholm, Vienna, Paris, London, the magnetic treatment has been successfully adopted in the cure of a variety of diseases. In epilepsy, paralysis, hysteria, neuralgia, chronic rheumatism, head-ache, I know of no remedy so immediate and availing. How often have I seen the victim of pain writhing in the most acute agony, sink under its influence into a state of the most placid composure! How often have I heard thanksgivings and prayers breathed in gratitude to the Creator for the relief which the afflicted have hereby experienced! Yes; and albeit the sceptic may smile incredulously,

pretend to perform great cures, began to try their skill to effect their recovery. Here it is necessary to observe that they use no medicines, either for internal or external complaints, but perform all their cures by charms." The learned traveller then describes somewhat ludicrously the processes they adopt; and adds, that however laughable they may appear in the eyes of an European, "custom makes it very indecent to turn anything of the kind into ridicule." *Vide* "A Journey from Prince of Wales' Fort, in Hudson's Bay, to the Northern Ocean. By Samuel Hearne. London."

he may rest assured that I have heard blessings which might move the heart of the coldest cynic, fall from the lips of those who under no other treatment could be relieved from the most intense and exquisite sufferings!

But it is not only in nervous diseases that animal magnetism is so valuable a therapeutical agent; in derangements of the vascular system, in the early stages of inflammation, and on the accession of fevers, it is equally useful. In cases of general debility, depending on scrofulous diathesis, to which young persons are so liable, it produces also a remarkably invigorating effect on the system. It ought, however, to be observed that the application of animal magnetism in the treatment of disease produces beneficial effects, albeit the more remarkable physical and psychical magnetic phenomena be not developed. The good, therefore, which is to be derived is not to be estimated by the ostensible magnetic effects which are induced.

"In some instances," says Loewe, "no other striking phenomena takes place, than merely a gradual change of the unhealthy to the healthy state, according to the repetition of the application. In others the magnetised person falls asleep, and is recovered without being brought to a higher degree of magnetical phenomena. The four following cases have come under my own observation:—At Gröningen, a girl, about nineteen years of age, was suffering under hysterical spasms, which sometimes continued for forty-eight hours: after being magnetised half an hour a day for three weeks, the patient recovered, and no other phenomena was effected.

" A lady residing in London, after a violent attack of fever, under which she was suffering in the months of December and January last, was affected by convulsions of every kind, but mostly by fainting, which often lasted two hours, and it was difficult to bring her to herself. I was one day present when this fainting was coming on, and, in presence of her sister and brother-in-law, I tried to make application of animal magnetism: I had scarcely begun to operate, when she quickly recovered from the fainting, as though she had been awakened by fright, and from that moment she gradually recovered, with the assistance of medicaments usual in such cases, and is now in perfect health.

" A boy in Gröningen, fourteen years of age, was suffering under tertian ague about eight weeks: after having given him all medicaments usual in this case, without effect, he was recovered by applying animal magnetism no longer than eight days; but no other phenomena was observed.

" A gentleman, thirty-five years of age, had laboured for many years under a chronic disease, which sometimes presented itself in the form of head-ache, particularly on one side of the head, (*hemicranium,*) and sometimes as an hypochondriac affection: after having made use of all the medicines usual in such diseases without effect, he was magnetised. For three weeks, during which time the animal magnetism was daily repeated, there was no remarkable difference in his case, but in the fourth week the symptoms of the first degree of magnetism presented themselves, and he was brought to the state of sleeping; and after the daily repetition of animal magnetism during the space of eight weeks, he recovered without any other phenomenon taking place."*

In other cases, however, the phenomena of clairvoyance and lucidity supervene, and the patients prescribe, not only for themselves, but for others, the treatment which should be adopted; and, as I have already advised, the injunctions given in this state should be scrupulously obeyed. I now proceed to adduce, from the mass of evidence which lies before me, a few cases which will sufficiently illustrate the curative effects of animal magnetism; I subjoin them without reference to any nosological arrangements.

Caroline Baudoin, twenty years of age, of a lymphatic temperament, had passed her childhood at Geneva, where the badness of her constitution fully developed itself, aggravated perhaps by the influence of the climate, or the use of unwholesome food. Her whole glandular system became diseased; her throat, breasts, and arm-pits exhibited tumours of a decidedly scrofulous nature, many of which suppurated, and discharged an abundance of purulent matter. The disease had been treated by all the most approved means; several issues had been inserted, but other tumours gathered and burst. One in particular, in the left arm, had burrowed down to the bone, and spread through the adjacent muscles, so as to render necessary the amputation of the arm. This having been resolved on, the operation was performed at the Hospital St.-Louis, with the consent of the patient, who, fatigued with the pain occasioned by this festering limb, looked upon its removal as a blessing. It was

* A Treatise on the Phenomena of Animal Magnetism, by M. Loewe. London. 1822.

attended with complete success ; the wound, however, took sometime in healing ; and the patient afterwards left the hospital, wherein she had been under treatment for several months. But her system, infected as it was with scrofulous matter, gave her no rest. Another wound opened on her breast, and resisted every medical treatment; she was in this state when I first knew her; a poor girl, doomed to great suffering, and apparently to a premature death. Moved by the recital of her sufferings, I resolved upon magnetising her, rather from an instinctive feeling that I might relieve her, than from any conviction that I could do her good, for I scarcely considered it possible to cure so inveterate a disease. In the course of three minutes' magnetisation, she fell asleep, and began by telling me that, had she known me seven months sooner, she would not have lost her arm. It was only three months since she had been operated upon. She pointed out the means of healing the wounds on the arm and breast, and on these being applied they proved completely successful. The most important thing, however, remained to be effected, which was to change her constitution, or at least to modify it in such a manner as to prevent a recurrence of the previous eruptions. Magnetism had produced a sufficient degree of lucidity to allow of her giving advice to other patients, but hitherto not enough to describe the means of curing herself. One day, as she was prescribing for a patient whose recovery she was anxious to bring about, she interrupted the consultation, and told me that on the 24th of August, at nine in the evening, she should fall into a state of profound sleep, which would last for thirty hours; that this sleep would be very calm, if during the two preceding days she were not annoyed by anything; but otherwise she should be much agitated; and that, by an unaccountable feeling, she should endeavour to eat her own flesh. She therefore desired that precautions might be taken to check this fatal propensity, and requested that she might be incessantly watched. She declared further, that during this crisis of thirty hours, she would eat absolutely nothing; and that the scrofulous matter would be carried out of her system. She also said, that during her sleep a *bruissement* would be heard at the epigastrium, caused by the flow of scrofulous humours. She then predicted her perfect recovery. This declaration was made on the 14th of July, 1833. I made her repeat it on the 21st of the same month, in the presence of fifteen persons, who drew up and signed a report to this effect, having previously taken care to ascertain her scrofulous state. In the intervening period many persons took cognizance of the declaration, and promised, if her prediction were fulfilled, to attest so remarkable a case. On the 24th of August, at eight in the evening, it was arranged that several persons should assemble in the house of the patient, at the Petit-Carreau; and I enjoined her attendants to put her to bed half-an-hour before the accession of her crisis, in order to prevent her being annoyed. All this was punctually done. At nine o'clock precisely, a number of visitors had congregated. On arriving, we were informed that the crisis had declared itself a few minutes sooner than she had predicted, and that it was fully developed. On entering the room we saw the unfortunate girl with her face swelled, her tongue protruding out of her mouth, nearly, to all appearance, cut in two by her teeth, her limbs stiffened, and her jaws so firmly locked that it was impossible to open them. After having magnetised the masseter muscles so as to remove the stiffness of the jaws, I caused the tongue to be drawn in, which was already very much discoloured, and fortunately had only been bitten very slightly. No one had yet perceived that one of her fingers had not only been bitten, but that there was a loss of substance, the piece wanting having been swallowed by her during her previous paroxysm. The wound was now dressed, out of which no blood, but a great quantity of red lymph issued. As the violence of this crisis continued, I thought it proper to remain with her during the ensuing thirty hours. I was perfectly right in having taken this resolution, for she struggled long with extraordinary violence, and attempted to put her hand into her mouth to bite it again, but she had been so bound down that she could only get at the sheets, a piece of which she succeeded in tearing off. The somnambulic state at length terminated; her prediction was fulfilled; and she was, to the satisfaction of all the parties interested, from that day cured.

A woman of forty years of age, worn out by long sufferings, and unable to stir without crutches, resolved on having recourse to animal magnetism, and was conveyed for that purpose to Paris, in a sedan-chair. She was two days travelling a distance of thirty-six miles. Several fainting fits came on during this journey. On being magnetised upon her arrival, she fell into somnambulism, but her sleep did not present any lucidity. The magnetic effects, however, produced by the magnetiser were such * that he at

* These effects can be easily imagined by any person who has seen a dead body galvanised. The patient, emaciated and extremely

once declared that, in a few days, the patient would walk without crutches. He therefore invited her to a ball, which he intended giving in his house on a certain day. The doubts expressed by the patient and her friends did not in the least disconcert him, but, on the contrary, he sent his invitations to a great many people, that they might go and convince themselves, and acknowledge the truth of his assertion, and the triumph of magnetism. In the meantime the patient was magnetised every morning, and at each sitting a decided improvement was observed. On the eleventh day she began to attempt a few steps, being supported by the arms, and she left off using crutches. On the seventeenth, she came to the *soirée* of the magnetiser, ascended the stairs unassisted, walked about the room, and remained there until one o'clock in the morning, when she retired to her own house without having felt any more fatigue than would have been the case with a person in good health.

Mademoiselle Lacourt, the daughter of a commissioner to the Mont de Piété, residing in Paris, Cour des Fermes, eighteen years of age, had suffered for five years and a half under morbus coxarius, affection of the hip joint; and had derived some benefit from the treatment she had undergone, when, on a sudden, in consequence of a fall, she lost the use of her legs; a luxation of the femoral bone now perfectly marked supervened. Able physicians were called in to attend her. The means prescribed by them were applied without any beneficial result, and they unanimously declared that the disease had assumed a scrofulous

pale, insensible, and immovable before magnetisation, suddenly, under the magnetic influence, starts up violently, and immediately the magnetisation ceases, falls again into a state of immobility.

character, which rendered it utterly incurable. This declaration was communicated to the relatives of the patient, and confirmed by M. Dupuytren, who advised that the leg should be amputated. Meanwhile another trial was made by applying moxas, which had been recommended by M. Broussais. The sufferer, however, obtained no relief. At last, some allusion was made to magnetism as a doubtful remedy, but the only one that remained untried. I was asked if I would employ it in this case, but I at first refused, considering that the disease was too severe, and most likely incurable; on being, however, again solicited, I yielded to the entreaty, and paid a visit to the patient, who did not even know what was meant by magnetism. After being magnetised by me for five minutes, she fell asleep; I then questioned her, and she declared, before several persons, that on the 25th of July, at twelve o'clock, she would be able to walk without crutches, that her lameness would leave her, and that she never should have a relapse. She requested to be magnetised daily for the next fifteen days, and then every other day up to the above date. This was five weeks previously, and I conformed entirely with her injunctions; and invited a number of physicians to watch the progress of her treatment. Twelve of my pupils observed it with great attention, and attended several times every week. The appointed day was approaching; the whole family, sadly afflicted by this dreadful calamity, could not believe in so prompt a recovery; it would have been a miracle—and nobody now-a-days believes in miracles—but I, in my own mind, was quite confident that the patient would walk. The effects produced by magnetism were very apparent at each successive sitting, and abundant exudation of viscous matter plainly indicated that a more healthy action was going on in the articulations; the skin recovered from its morbid colour; her strength reappeared; and her digestion decidedly improved. In fact, no bad symptom remained, excepting that the affected leg, as is usual in such cases, after having been longer, became shorter than the other. On the 24th, she again affirmed that on the morrow, at twelve, she would be able to walk without crutches, and without being lame. It was then six months since she had been confined to bed, and five years and a half since the first appearance of her disease. On the 25th, at half-past eleven, I went to her house; the apartment was crowded with persons wishing to witness this unprecedented cure. Incredulity was depicted on every countenance. I was called an enthusiast, a visionary, a mere youth; every one expected to witness my confusion. I confess that I felt an inexpressible anxiety; a cold dew trickled down my forehead; I was pale and trembling, notwithstanding something that whispered to me that in a few minutes I should reap the reward due to my exertions, and that my faith in somnambulic prevision, so often put to the test, would be once more justified. At twelve o'clock precisely, I approached the patient; she was in her bed, and dressed. I magnetised her before the spectators; and when sleep was induced, I reminded her of her promise, and called upon her to keep it; I then commanded her, in a firm voice, to rise and walk. She slowly moved out of bed, and placed her feet on the floor for her shoes to be put on. The spectators, impressed with the novelty of the scene, remained silent, staring, and motionless; then, at the waving of my hand, they drew back to make room for her; and I again commanded her to walk; she immediately touched the floor with her diseased foot, and very hesitatingly; she then made one step, took a deep breath, and at length walked forward to the other end of the room, without being supported or leaning on any one. She then returned to her bed. I now awakened her, and bid her again walk,—an act which she was perfectly unconscious of having just performed; she thereupon again touched the floor, to make sure of her footing, and moved with hesitation; but soon gaining strength, she walked several times round the room.

In the thesis of a distinguished foreigner, M. Albert Jozwik, which was cordially received by the professors,* who encouraged him to proceed in his scientific researches, several remarkable magnetical cases are reported. I extract the following,—"In the month of July, 1829, in the camp then before Warsaw, a subaltern officer of the third regiment of *Chasseurs à pied* of the Polish army, shot himself by putting the muzzle of his musket in his mouth. The medical officer of his regiment was instantly on the spot, and gave him every assistance, but in vain. This case was reported to me, as I was then superintending the medical department of the division. The body of the severely wounded officer had been conveyed to the infirmary, to which I immediately repaired; and, having found it still warm, I magnetised it. After a magnetisation of about half an hour, the poor fellow began to breathe, and was resuscitated; I then dressed

* MM. Pelletin, Dumeril, Andral, Rostan, and Chomel, were the examiners.

his wound, and sent him to the hospital called Uiazdow."*

M. Joseph Skrodzki, captain, aged 21, having been seized with an attack of St. Vitus' dance, was treated by the doctors of Besancon, who cauterised his arms and lumbar region, and bled him several times. Having found that the antiphlogistic treatment, and various medicaments prescribed in similar cases, had produced no change for the better, they advised him to take a trip to Paris. He accordingly came to the capital, and was admitted at the hospital de la Pitié, where he was again bled and subjected to applications of moxa at the nape of the neck. All these means were prescribed and strictly adhered to without producing any remarkable amelioration of health. The patient then applied to me, as he had before been under my care; but his state did not allow me to give him a favourable prognosis; I proposed, however, to treat him magnetically for a fortnight, that I might then be enabled to pronounce whether it was in my power to cure him. He had left the hospital de la Pitié in the month of July, 1832. I began to apply magnetism immediately on his leaving. At the first sitting, the paroxysms which had occurred in the early part of the day no longer returned; they were confined to the *prodromes;* during the subsequent days, the prodromes were less frequent. On the seventh day the captain was perfectly cured, and the parts which were in a purulent state were healed by a lotion of magnetised water.

* This case clearly proves that, as a *resuscitating* process, animal magnetism may, in cases of partial asphyxia, be advantageously applied.

A child, eight years old, without any known cause, was seized with ascitis, and afterwards with an ulcer in the right leg. The physician who attended wished to perform the operation of paracentesis, and, with this view, applied to me for advice. I proposed to defer the operation for a few days, and to submit the child to the magnetic treatment, to which they acquiesced. As early as the second day of magnetisation, the evacuations became more abundant than they had been during the application of the remedies, which had been previously exhibited with this intent. After the third, fourth, and fifth sittings, the patient became still more improved, and by the twentieth day was perfectly cured.

The curative influence of animal magnetism was fully attested by the Report of the French Academy, to which I have so often adverted—not because the facts it records are magnetically considered curious or remarkable, but because this is an authentic document, embodying the verdict of a critical tribunal, which herein put on record details which, as pure matter-of-fact evidence, cannot be controverted. Any individual authority, however unexceptionable, is liable to be impugned; but I do not understand how the concurrent testimony of numerous scientific men, highly respected members of one of the most scientific academies in Europe, who were delegated, in the most solemn manner, to inquire into the veracity of any series of demonstrable facts, can be so easily repudiated. On this account, I repeat, I so frequently cite this report in preference to my own experience, or any other authority. The following case, of an old and severe paralysis cured by animal magnetism, is very similar in its details to some of those above narrated.

"Paul Villagrand, student of law, born at Magnac Laval (Upper Vienne,) on the 18th of May, 1803, suffered a stroke of apoplexy on the 25th of December, 1825, which was followed by a paralysis of the whole left side of the body. After seventeen months of different modes of treatment, by acupuncture, a seton in the nape of the neck, twelve applications of *moxa* along the vertebral column—modes of treatment which he followed at home, at the Maison de Santé, and at the Hospice de Perfectionnement, and in the course of which he had two fresh attacks—he was admitted into the Hôpital de la Charité on the 8th of April, 1827. Although he had experienced perceptible relief from the means employed before he entered this hospital, he still walked with crutches, being unable to support himself upon the left foot. The arm of the same side, indeed, could perform several motions; but Paul could not lift it to his head. He scarcely saw with his right eye, and was very hard of hearing with both ears. In this state he was entrusted to the care of our colleague, M. Fouquier, who, besides the very evident paralysis, discovered in him the symptoms of hypertrophy of the heart.

"During five months, he administered to him the alcoholic extract of *nux vomica*, bled him from time to time, purged him, and applied blisters. The left arm recovered a little strength; the head-aches, to which he was subject, disappeared; and his health continued stationary until the 29th of August, 1827, when he was magnetised for the first time by M. Foissac, by order and under the direction of M. Fouquier. At this first sitting, he experienced a sensation of general heat, then twitchings (*soubresauts*) of the tendons. He was astonished to find himself overcome by the desire of sleeping; he rubbed his eyes in order to get rid of it, made visible and ineffectual efforts to keep his eyelids open, and at length his head fell down upon his breast, and he fell asleep. From this period his deafness and head-aches disappeared. It was not until the ninth sitting that his sleep became profound; and at the tenth he answered, by inarticulate sounds, the questions which were addressed to him. At a later period, he announced that he could not be cured but by means of magnetism, and he prescribed for himself a continuation of the pills composed of the extract of *nux vomica*, sinapisms, and baths of Bareges. Upon the 25th of September your committee repaired to the Hôpital de la Charité, made the patient be undressed, and ascertained that the inferior left limb was manifestly thinner than the right—that the right hand closed much more strongly than the left—that the tongue, when drawn out of the mouth, was carried towards the right commissure, and that the right cheek was more convex than the left.

"Paul was then magnetised, and soon placed in a state of somnambulism. He recapitulated what related to his treatment, and prescribed that, on that same day, a sinapism should be applied to each of his legs for an hour and a half; that next day he should take a bath of Bareges; and that, upon coming out of the bath, sinapisms should be again applied during twelve hours without interruption, sometimes to one place, and sometimes to another; that, upon the following day, after having taken a second bath of Bareges, blood should be drawn from his right arm to the extent of a *palette* and a half. Finally, he added, that by following this treatment he would be enabled, upon the 28th, *i. e.*, three days afterwards, to walk without crutches on leaving the sitting, at which, he said, it would still be necessary to magnetise him. The treatment which he had prescribed was followed; and upon the day named, the 28th of September, the committee repaired to the Hôpital de la Charité. Paul came, supported on his crutches, into the consulting-room, where he was magnetised, as usual, and placed in a state of somnambulism. In this state, he assured us that he should return to bed without the use of his crutches, without support. Upon awaking, he asked for his crutches; we told him that he had no longer any need of them. In fact, he rose, supported himself on the paralysed leg, passed through the crowd who followed him, descended the step of the *chambre d'expériences*, crossed the second court of the building, ascended two steps, and when he arrived at the bottom of the stair he sat down. After resting two minutes, he ascended, with the assistance of an arm and the balustrade, the twenty-four steps of the stair which led to the room where he slept, went to bed without support, sat down again for a moment, and then took another walk in the room, to the great astonishment of all the other patients, who, until then, had seen him constantly confined to bed. From this day Paul never resumed his crutches."

But however efficient a remedy animal magnetism may be in paralysis, it is still more so in cases of epilepsy;* and all medical men are aware that this is one

* There is a certain analogy between the magnetic and epileptic state of insensibility. It has been observed, that "when the patient awakes from a fit of epilepsy he has generally no recol-

of the most intractable diseases that can fall under their attention. Its pathology is yet very imperfectly understood; and its treatment, on the ordinary principles of medical science, is generally a confessed failure. In vain has the physician exhausted his ingenuity, and ransacked the vegetable and mineral kingdoms to discover some remedy which might avert the paroxysm, or mitigate its intensity. Before the perplexed bystanders, the individual suddenly falls down in a state

lection of what has passed; and perhaps, therefore, there is no suffering. The want of recollection of suffering is no proof that there has been no suffering; for we have all suffered enough in cutting our teeth, and we know nothing of it now. And so it may happen in more recent events; the fit may be attended with more or less suffering, and yet the individual not be aware of it afterwards." "But," says Dr. Elliotson, "I should think there is no suffering; and for this reason: persons do not suffer, in general, when they are hung. There is an account in Lord Bacon's works of a person who was hung, and all but killed, and yet he did not suffer. There is a short account of Cowley, the poet, (which is very scarce,) from which it appears that he three times attempted to commit suicide, and one of these attempts was by suspension. The account was written by himself, and found among his manuscripts. He there mentions, that he suspended himself over the chamber door in the Temple, and became perfectly insensible. He only recollected a flash of light appearing before his eyes. His weight at last caused him to drop on the floor; there he was found, and, after a time, recovered. He says, that although he was thus in the jaws of death, and had become perfectly insensible, yet he had no previous suffering; and, therefore, as there was no previous suffering in that state, it is probable that there is no suffering in epilepsy. I should suppose, that in drowning there is no suffering, if it occur at once. Shakspear's expression is, 'O Lord! methought what pain it was to drown.' But there is no reason to suppose there is pain, if the individual go down and do not come up again; but if he come out of the water, the suffering is dreadful."—Good's Study of Medicine, by S. Cooper, vol. iii., p. 410, *note.*

of perfect insensibility ; the limbs become convulsed, the body writhes in apparent agony, and sometimes doubles on itself (*oposthotonos*) ; the jaws are clenched, the teeth gnash violently on one another, and the tongue protruding between them is often severely wounded. It is no wonder that, in a superstitious age, it should have been supposed by ignorant persons that the victims of this strange malady were possessed by an evil spirit, and that they sought the exorcism of the priest rather than the aid of the physician. We may rest assured, however, that during the mystic rites of exorcism the priests, as I shall elsewhere have occasion to shew, acted magnetically; under the form of a religious ceremony they communicated the magnetic influence, and in consequence of the beneficial results they produced, these rites are still preserved in many catholic countries. Hence a celebrated English writer, the late Mr. Thomas Hope, the accomplished author of Anastasius, observes, " there are certain nervous complaints, such as epilepsy and convulsions, at once violent while they last, and during their intermittence leaving no trace behind. Of these, the infliction appears to proceed more immediately from superhuman causes, and the cure to be more directly within the province of the priesthood, and are only to be removed by exorcisms and prayers." He then adds—and I beg attention to that which follows—" I myself have witnessed at Naples the ceremony of expelling evil spirits; and, strange to say, they yielded for a time to the imprecations pronounced against them."* It is certain that no ordinary medical treat-

* Essay on the Origin and Prospects of Man, by Thomas Hope. 3 vols. John Murray, Albemarle street. Vol. III., pp. 148, 149. The evidence of Mr. Hope is, perhaps, the more valuable, as the

ment is of any avail in this disease. The reason, says Dr. Elliotson, it is so generally intractable, and that so many remedies are so uncertain and so unsatisfactory, is very evident. This is a disease which arises from every sort of irritation in every part of the body; and the irritation may be structural, may be slow inflammation, or something we cannot remove. If it arose from one cause, it would be a different thing; but it will arise from any cause whatever, physical or mental, organic or inorganic, and situated in any part of the body.* Now, reasoning pathologically, the causes of epilepsy as here described, and as given, also, by numerous French authors, are precisely those in which it might be predicated that the application of animal magnetism would prove availing. I know of no remedy so likely to allay irritation, from whatever cause it may arise; and I now speak practically, not theoretically, for I have myself succeeded in curing many cases of epilepsy by magnetism, and so have other magnetisers. Unless, indeed, this disease depend on organic lesion, or some purely structural cause, I consider epilepsy perfectly curable by magnetism; and, even in those cases, the intervals between the paroxysms may, by the magnetic operation, be prolonged, and the intensity of the paroxysm itself mitigated. I do not, be it observed, pretend that animal magnetism will cure all diseases; I do not cry it up as an universal remedy or *panacea*, which is at once to exterminate

whole tenour of this work is to establish the doctrines of materialism; and, as one of the sceptical philosophers of Hume's school, he would naturally be cautious how he admitted the existence of any facts contrary to the *ordinary* routine of *experience!*

* Lectures delivered at the London University. Medical Gazette, 1833-4.

suffering, and prolong the duration of human life. Far from it; I regard it as an important and very valuable therapeutical agent, the exhibition of which requires much skill and judgment on the part of the medical practitioner. Nay, there are many diseases in which I have learned, from experience, that its application is decidedly *contra-indicated*, and would be attended with injurious consequences. Already I have stated, that, in the case of a female patient who was magnetised by me at the Hôtel-Dieu of Paris, the pulse rose from 60 in the minute to upwards of 100, an increase of circulation which might be attended with a fatal result in a person labouring under aneurism, or disposed to hæmoptysis, (hemorrhage of blood from the lungs,) in far advanced pulmonary consumption. When I first commenced the practice of magnetism, elated with the success I had obtained in some very inveterate and almost hopeless cases, I magnetised patients indiscriminately, and, among others, several afflicted with pulmonary complaints, which other physicians had pronounced incurable. It appeared to me —I was at any rate sanguine enough to believe—that consumption itself would yield to its beneficial influence, for I had read that it had in some cases been successfully applied;—however, I was deceived, as the following results may testify. A patient suffering under this complaint being magnetised by me, at first experienced a general soothing effect—a calmness, accompanied by a sensation of general comfort. But this state was soon succeeded by violent paroxysms of coughing, attended by profuse perspiration, and acute pains in the chest. Thus affected, she earnestly entreated me to desist; but, supposing that some favourable crisis was at hand, I persevered, and found these ominous symptoms only became aggravated; the cheeks became flushed, the eyes suffused, and the convulsive coughing brought up a quantity of bloody mucus. I was then compelled to suspend magnetisation; and, from further experience, have satisfied myself that in such cases its application is inadmissible.

"Philanthropy, or the desire of relieving a fellow-creature," says an intelligent author, "has, no doubt, greatly magnified the real virtues of magnetism; but this power indubitably exists; it is, therefore, the duty of the physician to study it without prejudice, and of the physiologist to circumscribe it within proper limits. The direct influence of this new agent on the nervous system leads us to think, that its power may be most efficacious in nervous affections in general. Hysteria, hypochondria, melancholy, catalepsy, epilepsy, have derived the most salutary benefit from its application; spasmodic diseases of every kind, cramps of the muscles, convulsions, and a multitude of pains, as well as cases of deafness and paralysis, have also been relieved by its application. In these various affections, the nervous system being the chief seat of the disease, and magnetism directing its influence especially on this system, it evidently follows that the result must be worthy of attention. It is, therefore, in these affections that the partisans of magnetism testify to having obtained the most surprising success. It would take up too much time were I to adduce instances of this; but every work on magnetism is replete with them. There is no *panacea* in nature, and we do not pretend to represent magnetism as an universal remedy; for if it be useful in some cases, it is nevertheless injurious in others. We must carefully, therefore, study the nature of its operation, and learn whether it be exciting, debilitating, or sedative; and having once succeeded in determining the precise nature of its physiological action, we may then proceed, in a philosophical spirit, to ascertain the particular cases in which it may be employed with benefit."

In many acute diseases, animal magnetism alone is not to be trusted, and certain medicines should at the same time be administered; but I shall abstain from entering into details which would be purely professional. However, I may here observe, that a good work on the practice of medicine, founded on the principles of animal magnetism, is much required by medical men; and this *desideratum* I propose endeavouring to supply in my next volume, to which this, as I have already stated, is only an introduction, which I now publish for the purpose of calling the attention of the profession, and the public generally, to the elementary facts of animal magnetism, before proceeding to more complex details; and this is the only course I could with propriety have adopted; for there is, in this country, a general disbelief in animal magnetism, accompanied by an indisposition even to investigate the facts; and if I am only fortunate enough to disturb this apathy, and excite even an interest in the subject, I shall esteem myself happy; for they who will conscientiously enter on the inquiry, will find such a multitude of facts spring up around them, that I am persuaded they will eventually be convinced of the truth I advocate. It is the sceptic only who will not inquire; it is he who hath eyes to see, but will not see,—who hath ears to hear, but will not hear; it is he only I am inclined to pronounce incurable; for, what would avail a revelation from

Heaven itself to so perverse a being? And is not all Nature, in reality, a revelation from Heaven to the mind of the philosopher who should walk along her paths rejoicing in the freedom of his thoughts, unencumbered by the fetters of those narrow prejudices which must fatally restrict the sphere of our comprehension?

CHAPTER VII.

POWER OF ANIMAL MAGNETISM—MAGNETISM OF INANIMATE OBJECTS.

THE penetrating intensity and subtile power of the magnetic influence is best illustrated by the magnetiser performing the operation at a distance, and without the knowledge of the patient, through the intervention of such opaque bodies as a screen, folding doors, or thick walls. The possibility of the operation being conducted efficiently in this manner is established beyond a doubt; and I proceed, therefore, to cite some few facts in proof of it, because, before entering on the explication of the phenomena of animal magnetism, it is necessary that all the facts should be distinctly detailed. On the 4th of November, 1820, the Commissioners having assembled at the Hotel Dieu, M. Husson, one of the physicians of the hospital, said to me, "You can induce sleep in a very short time; I wish you would obtain the same result without the patient either seeing you or being aware of your presence." I replied that I was willing to try, but that did not warrant the success of the experiment, because the action at a distance depended so much on the peculiar susceptibility of the individual. We agreed, however, upon a signal which I could hear in an adjoining apartment. M. Husson was to throw a pair of scissars down upon the table, and I was then to begin the operation. Accordingly I was ushered into a closet, separated from the room by a thick partition; and the door having a good lock and key, I felt no hesitation in locking myself in, as I wished to multiply obstacles, so that no pretext might be left for suspicion or doubt. The patient was then brought into the adjacent room, and seated at a distance of about three or four feet, with her back turned towards the closet wherein I was concealed. They then, to deceive her, began wondering why I had not yet arrived, and they pretended from this delay to conjecture that I did not intend to come; at the same time, they blamed me for not keeping my appointment. In short, my alleged absence was adverted to in such a manner as to leave no doubt, on the mind of the patient, that I was really elsewhere. The signal being then given, although I knew not where nor at what distance Mlle. Samson was placed, I began to magnetise in the most profound silence, carefully avoiding the least movement which might apprise her of my being present. It was then thirty-five minutes past nine. In the course of three minutes, as soon as my will began to act, she was observed to rub her eyes, exhibit the usual symptoms of approaching sleep, and at length fell into a state of somnambulism. This experiment I repeated on the 7th of November, before Professor Recamier, who took every precaution to prevent deception, and the result was precisely similar. The following are the details of this second trial:—On my entering, at a quarter past nine, the operating room, M. Husson informed me that M. Recamier wished to be present, and to see me put the patient to sleep through the partition. I readily acquiesced. M. Recamier then entered, and privately conversed with me on the subject, when we agreed upon a signal; and I slipped into the closet, where I was again locked in. Mlle. Samson was then brought in; M. Recamier placed her with her back turned towards the closet, at a distance of more than six feet from it. He then proceeded to converse with her, and found her better; but on being told that I was not likely to arrive, she insisted upon retiring. The moment M. Recamier asked her whether she *digested meat*, which was the signal agreed upon between us, I began to magnetise her at thirty-two minutes past nine. Three minutes after, M. Recamier touched her, raised her eyelids, shook her by the hands, put questions to her, pinched her, and convinced himself that she was in a state of profound magnetic sleep. But even these trials were thought insufficient to establish so extraordinary a phenomenon; and it was resolved, therefore, to multiply them, and vary all their details. Accordingly, I went one night, accompanied by M. Husson and other physicians, to the ward in which the same patient was, and took my position from her at a distance of several beds, keeping at the same time the most profound silence, so as not to attract attention. I began magnetising her at eight minutes past seven. At twelve minutes past, we all drew near, and convinced ourselves that magnetic sleep had been induced. It is unnecessary to add, that the period fixed on for this operation had been selected, not by me, but by the principal physician of the hospital; further, that it had been ascertained that the patient was not asleep before the experiment, and that the magnetic influence had been exerted at a distance of about twenty feet. Having witnessed these experiments, M. Bertrand, Doctor of Medicine of the University of Paris, affirmed that he did not think it extraordinary that the patient should fall asleep while the magnetiser was in the closet; because the constant recurrence of the same attending circumstances might have, independently of my presence, produced the same results; and that, moreover, the patient might be naturally predisposed to sleep. He therefore proposed that she should be brought into the same room, and seated on the same seat, and that a conversation should be held with her as before. This repetition of the same circumstances, he thought, would induce sleep. But he was deceived. At a quarter before ten, they began to execute the plan agreed upon;—she was seated on the same easy chair where she usually sat, and placed in the same position. Various questions were put to her; she was then left to her own thoughts; the signals previously used were again repeated,—scissars were thrown upon the table; they went, in fact, through a complete rehearsal of everything previously done, but in vain was the magnetic state expected. She complained, became restless, shifted her position, and evinced anything but a disposition to either natural or magnetic sleep. After this, on my entering the room and commencing the operation, the magnetic sleep, accompanied by the usual state of physical insensibility was perfectly established.

In a letter, dated Lyons, 1784, which appeared in a work entitled, Reflexions Impartiales sur le Magnétisme, it is stated:—"The following experiment has often been repeated. A very susceptible somnambula was left in a room with other persons endeavouring to engage her attention, during which time she was, without her knowledge, magnetised from the next room, and the effect was as prompt and nearly as powerful as if the magnetiser had been close to her; the only difference observable was that, not being aware that she was under the magnetic influence, she at first endeavoured to counteract the sensations she experienced, mistaking what she felt for a natural uneasiness, and she only gave way to it when the magnetic action, increasing in power, forced upon her the consciousness that she was being magnetised. A single experiment would not have been decisive; but by repeating the operation results have been obtained more or less characterised according to the degree of sensibility in the person magnetised." The report of the "Académie de Médecine" also contains a case of magnetic action at a distance, which was proved to the conviction of the commissioners—"On the 10th of September, 1827, at seven o'clock in the evening, the commissioners," says the reporter, "met at the residence of M. Itard, who continued his experiments on Cazot. The latter was in the room where a conversation was kept up with him till half-past seven, when M. Foissac, who had entered after him, and remained in a parlour separated from the above room by two doors well closed, and at a distance of twelve feet, began to magnetise him. Three minutes after, Cazot said:—'I think M. Foissac is there, for I feel quite giddy.' In the course of eight minutes he was in a profound sleep."

It has already been observed that females can perform the magnetic operation as well as men; and the following case, related by M. Henin, will, I doubt not, be perused with interest:—"An Indian lady," says he, " residing in Paris, and possessed of great magnetic power, was solicited by a lady who accompanied me, to visit her, and give us a proof of the energy of her will. She had a maid servant, whom she kept under the magnetic treatment, and often threw into somnambulism. She was then working in a room entirely separated from us. The Indian lady, on being asked whether her maid should appear before us at her tacit mental bidding, composed herself for a moment, and magnetised her from the room where she then was sitting without speaking or making the slightest motion. A few minutes afterwards, we saw the maid-servant step into the room in which we were to inquire of her mistress what was her pleasure." M. Henin then adds, " that it was not possible he could be deceived, and that he would not have related the fact had he not been convinced by the clearest evidence of its existence." And after all, is there anything more astonishing in the transmission of the animal magnetic fluid through intervening opaque bodies—than in the transmission of the galvanic or electric fluid which we know permeates matter with incredible velocity? Here, again, the wonder is not greater in the one case than it is in the other; albeit the one excites incredulity while the other is universally admited. The will, I am convinced, only acts in these cases as the *primum mobile*, which directs, under its energies, the direction of the magnetic fluid, which is an animalized principle that may even be accumulated in inanimate objects, and this, I doubt not, was the great secret of the amulets and talismans which were by the ancients supposed to possess medicinal properties. Hence, Galen himself is reported to have cured epilepsy by suspending peony roots round the necks of his epileptic patients; and the cures which he effected by magnetic processes gave him the reputation of a wizard, and compelled him to quit the Roman capital. A kind of grateful veneration for these inanimate objects, on account of the virtues they were presumed to possess, transformed them by degrees into tutelary divinities, and superstition, which ever walks in darkness, then extended similar virtues to all sorts of ridiculous objects, which were condemned as dangerous. Hence the Council of Laodicea; in the fourth century, prohibited the use of amulets, and this prohibition extending to rings of all descriptions was repeated and confirmed by the Council of Rome in 712—of Milan in 1565—and of Tours in 1583. The discovery of animal magnetism, and its investigation on scientific principles, throws considerable light on a variety of facts which, in the darker ages, led to horrible persecutions, and surely, if history can bring any moral home to our conviction, it should teach us how abject a being is man, when, armed with the fire-brand and the sword, he puts himself in an attitude of hostility before the eternal divinity of truth.

In accordance with the views above stated, we shall find that the animal magnetic fluid, or that invisible principle or agent which produces the magnetic phenomena, pervades all bodies,—in other words, all are conductors of this fluid. It may be incorporated or accumulated in them so that they may receive and retain it, and thereby exhibit magnetic phenomena, and the affinity between the animal-magnetic fluid and these bodies into which it enters is so great that no chemical or physical power can separate them. Chemical re-agents and caloric have no power over it. There is, however, very little analogy between the imponderable fluids known in chemistry, and the animal-magnetic fluid. In proof of all which I shall here refer to numerous experiments which are well authenticated. A magnetised vitreous body, which had put a somnambulist to sleep in a few seconds, was afterwards rinsed with water and wiped with a linen cloth; on being again presented to the same magnetic subject he fell asleep in one minute and a half.

The same magnetised glass, rinsed with alcohol, produced sleep in half a minute.

Another magnetised glass, rinsed with ammonia, elicited somnambulism in fifteen seconds.

The same glass was plunged into fuming nitric acid. After an immersion of five minutes it was put into a China cup with water, out of which the young somnambulist having taken it, fell immediately asleep.

The same experiment was repeated with concentrated sulphuric acid, and the result was exactly the same.

In all these experiments no chemical re-agent could destroy the magnetic power of the magnetised glass. Hence it follows that this power, unlike colours, electricity, and other similar fluids, does not reside merely at the surface, but that it penetrates the whole mass. The following experiment seems to confirm this natural conclusion:—

A large marble pestle, after being magnetised, was wholly immersed in muriatic acid, in which it was kept until the acid had reduced the mass to about one-half of its original size, it was then drawn out, well cleansed, and presented to the somnambulist, who fell asleep as soon as he had touched the whole mass of marble.

Other imponderable and expansible fluids bear a certain relation to each other; namely, when one of them adheres to a solid body it cannot be separated from this solid excepting by the action of some other expansible fluid. For instance, the mineral magnetic fluid is expelled from a natural or artificial loadstone by ignition, and by an electrical shock. Combustion, indeed, alters all relations between solid bodies and expansible fluids, such as their conductive properties, capacity, &c. Hence, in order to investigate the nature of the animal magnetic fluid, it was absolutely indispensable to submit the magnetised body to various degrees of temperature, and even to combustion. Accordingly, magnetised wax, colophane, sulphur, and tin, were melted, and then poured into cylindrical moulds similar to their original shapes. Their virtue was then tried upon the somnambulist without the slightest difference in the results before and after this operation: the instant he laid hold of them he fell asleep.

A magnetised iron rod was made red hot, and in that state plunged into a cup full of water, and then presented to a young man by the same person who had performed the operation, and as soon as he held it in his hands he fell asleep.

A large sheet of paper, twisted and magnetised, was burnt in a *faïence* plate; the carbon and cinders which remained were presented to the somnambulist, who took up as much as his hand could hold, and fell asleep in a few moments.

Many cross experiments were tried with objects

which were not magnetised; but no effect whatever by them were produced; but those which were magnetised being preserved with care, produced the same effects six months afterwards;—they seemed to have lost none of their magnetic power.

All these experiments, made with the greatest care by Professor Reuss and Dr. Lœwenthal, two physicians residing in Moscow, have been repeated by me with hardly any difference in the results. I have, moreover, discovered that whenever a somnambulist, whether asleep or awake, approaches a spot where magnetised objects are concealed, his physical organisation undergoes remarkable modifications. In most of my experiments, I had taken especial care to blindfold the somnambulists; and in no instance did I ever communicate to them the particular object I had in view, without observing this effect. One day, after having experimented, and tried with great perseverance, whether some natural bodies had not the property to isolate the somnambulist from the magnetic action, I fell upon the device of taking, out of the library of the gentleman in whose house I was experimenting, a large and voluminous folio: I magnetised the somnambulist through the covers of the book, holding my fingers in a pointed direction. He soon experienced all the effects of magnetisation, as powerfully as if no body had been interposed between him and me. I then endeavoured to magnetise the somnambulist through the whole book; and soon perceived, with some surprise, that the epigastrium, which I tried to operate upon, felt nothing, whilst the head and feet visibly experienced the magnetic action. It was only after persisting for some time that the epigastrium also felt the agency of magnetism; but still the effect was much weaker. I repeated this experiment on other organs, and they were very slightly influenced, while the rest of the body was actually convulsed. Experiments, repeated under every variety of circumstances, do not allow me to doubt this extraordinary fact,—namely, that of all known bodies, the best calculated to intercept the transmission of the magnetic fluid is a great number of super-imposed leaves of paper, as in a book.

"Hence it follows," says M. de Eschenmeier, a celebrated philosopher of Tubingen, "that there is an active principle resisting every mechanical, physical, and chemical power, adhering to all solid bodies by an indissoluble bond, which pervades their substance like a spiritual being, and sets at defiance even the devouring action of fire. But its existence, made manifest by the phenomena it produces, does not reveal itself to the organs of sense in the usual state of man; but only during this delightful expansion of our own individuality, effected by the magnetic relation, which bestows upon us the faculty of seeing, hearing, and feeling, this vital principle, which receives all its vigour from the will of man, and operates with an energy proportionate to the strength of the will. When it acts with great energy upon an organ endowed with an equal but negative power, which supposes the existence of a specific contrast (as when a strong man directs it on a weak boy,) this principle has all the intensity of lightning, and seems to extinguish life at one stroke. In the usual waking state, man is but in a general relation with surrounding objects; in this magnetic state, by the sole energy of his will, he defends his personal individuality against all extraneous influence which threatens the spiritual part of his existencc, and this will is more or less in equilibrium with the will and action of other creatures. But this resistance exists only as long as the body and the soul preserve their intimate connexion. In this state we possess a perfect consciousness of ourselves; and all our notions and sensations, as well as our will, being in harmony with the welfare of the body, preserve between them a just equilibrium and equal proportion. In this state, which may be considered as intermediate between the purely spiritual state and that of the brute creation, man sees before him, on the one hand, an ideal world, and, on the other hand, a corporeal world. But as long as his own individuality remains firm, and he preserves the consciousness of his own being, he can never enter either of these worlds. He can only elevate his thought by giving to all his sentiments and actions the stamp of truth, of beauty, and of goodness; but he is nevertheless not able to soar into that region where thought enters into a state of transcendent purity. It is only when freed from all the shackles of its material mould that it can reach this sublime height. These are the two extreme limits which bound the range of man's existence during the habitual waking state. The existence of a new specific relation can change this state. A foreign impulse, particularly the will of a stranger, can penetrate this circle of indifference which determines the ordinary state; it can open all its barriers, and overcome its obstacles, to a certain extent. On the one hand, the spiritual essence of man becomes more refined; it is no longer confined to its former seat; it no longer contemplates the realms of thought as a remote constellation; but, bounding at once over the limits of the senses, it acquires new perceptions on entering a new state. On the other hand, the organic system becomes more materialised, begins to act as the imponderable fluids of electricity, mineral magnetism, &c. In this state of exalted contrast, man is capable of perceiving the vital principle itself, of seeing, hearing, and feeling it as a distinct object; but in his ordinary condition, this is utterly impossible, since it is the same vital principle which itself actively hears, sees, and feels; it cannot, therefore, be passively heard, seen, or felt. In the magnetic waking state, when the contrasts are exalted, the spiritual essence is less closely united to the organic system, the eye of fancy itself becomes more intelligent, raises itself above the vital principle, and perceives it as an object. We cannot explain in clearer terms the constant action of magnetised substances upon the somnambulists, and the immediate influence of magnetism. In magnetic vigilation, the spiritual essence is no longer under the dominion of the vital principle; it looks upon the latter as upon a subordinate being. In like manner the somnambulist, being brought under the energetic influence of the magnetiser's will, can send forth this vital principle as a messenger to reconnoitre the most remote regions. His eye, like a ray of light, darts forward to an immeasurable distance, taking no other direction but that prescribed by the will of the magnetiser. It is the same vital principle which adheres to all bodies by an indissoluble combination, without being perceived in the usual waking state, because this state admits of no specific relation, and is rather opposed to all influences; but in magnetic vigilation it is distinctly felt by a perceptive organ."

Assuredly no person, however cynical, can condemn an attempt tending to unfold the mysteries of nature to our contemplation,—mysteries which are constantly reproduced before us; for what is the formation of bodies, their motion, the origin and cure of diseases, but ineffable mysteries? Do we understand the connexion between the soul and her corporeal tabernacle, which is ever vibrating in unison with the infinite modulations of her thoughts, and the exquisite harmony of her feelings? No. Let proud philosophy, therefore, descend from her throne of bigotry and intolerance, and with a spirit of humility proportioned to their sublimity, prepare to investigate these solemn mysteries. It is, I repeat, the mystery of life, and the not less absorbing and perplexing mystery of death; it is the revealing to us of that spiritual essence upon which are shadowed the outward forms of all material objects and living beings; it is the unveiling to us of an hereafter which all men anxiously desire, yet which the *dicta* of materialism would annihilate. I know of no study so exalting to the human mind, none so deeply interesting to all who are capable of reflecting on their own destiny.

CHAPTER VIII.

TESTIMONY OF THE ANCIENTS.

ALTHOUGH animal magnetism may be regarded as a modern science, the principle upon which it is founded, and the phenomena it educes, were recognised at an early period. Most of the magnetical facts I have above described were familiar to the ancients. It should be remembered that a long time is required before any science can attain maturity. Ages must elapse during which isolated facts only can be observed, and these will remain scattered like the links of an unconnected chain, until some master spirit rises up which can discriminate the affinities that subsist between them, and determine the laws by which they should be mutually associated. Hence vestiges of almost every science may be traced in the early history of mankind ; but the facts are often strangely confounded with, and obscured by, the prevailing superstitions of the age. It is, however, for a more enlightened philosophy to divest truth of all accessory fictions, that her shrine may be revealed to us unclouded by the mists of prejudice, and the illusions of imagination. It is not surprising that so universal a power as that of animal magnetism, producing effects so ostensible, should have been recognised even in the earliest ages. The ancients were well acquainted with the phenomena of somnambulism; they discovered that it could be induced by certain mystic rites, during which, as will hereafter appear manifest, the magnetic principle unknown to them was evolved; and they placed implicit reliance on the lucidity which occurs in this state, and which they naturally enough ascribed to the direct interposition of some of their heathen divinities. We shall find, too, that they had recourse during these religious ceremonies to direct manipulations; hence Prosper Alpinus distinctly states, that frictions during ablution was one of the secret remedies of the Egyptian priests,* after which the patients lay down on the skins of the animals they had sacrificed, and, having prepared themselves for sleep by fervent prayer, awaited the visions which Isis was supposed to reveal to them. "The Egyptians," says Diodorus Siculus, "report that Isis found out many medicines for the recovery of men's health: being very expert in the art of physic, she contrived many remedies for that purpose, so that even now when she is advanced to an immortal state, she takes pleasure in curing men's bodies. For clear proof of all this they refer, not only to the usual fables of the Greeks, but to the undoubted evidence of the fact......... Many who have been given up by the physician as incurable, have by her been recovered; nay, many who were perfectly blind, have, by her aid, been restored to sight and soundness of body."† The same ceremony was observed, according to Strabo, in the temple of Serapis,‡ and,

* Usus frictionum in balneis apud Ægyptios ita est familiaris ut nemo ex balneo non frictus abeat.—Prosp. Alpin. de Medic. Ægypt. lib. iii. c. 18.

† Diodorus Siculus, lib. i. c. 25.

‡ Strabo, lib. xvii.

on the authority of Galen, in the temple of Vulcan, called *Hephestium*, near Memphis.* In the temples of Apollo and Æsculapius, ceremonies were conducted on the same principle, the object of all, as interpreted by Sprengel, being to occupy the imagination of the patient, and produce a state of exaltation sufficient to elicit the desired effect.† He also observes, that the baths were always accompanied by frictions and other manipulations, which could not fail to produce surprising effects on persons possessing a delicate nervous system.‡ When cures had been thus successfully wrought, it was customary to record them on tablets of marble or brass, which were erected in the temples; and sometimes they were engraven on the walls or pillars of the temple itself. Many of these inscriptions have been lost, but some few have been preserved by the antiquarian researches of Mercurialis, Gruter, and others. They constituted the only code of medicine that existed at this epoch; and it is even affirmed that the pretended works of Æsculapius were borrowed from these inscriptions in the temples of Cos.§ "Nothing," says Dacier, "is more common among the ancients than remedies indicated to the sick in their dreams; and this was so generally received in antiquity, that they were sought in the temples, because it was believed that the god more readily com-

* Galen, lib. v. De Med. Sect. genes. c. 1.

† Le culte rendu à Esculape avait pour but d'occuper l'imagination des malades par les cérémonies dont ils étaient témoins, et de l'exalter assez pour produire l'effet que l'on désirait.—Histoire de la Médecine, par K. Sprengel, tom. i. p. 144, ed. 1815.

‡ Ils étaient accompagnés de frictions et *autres manipulations* qui devaient opérer des effets surprenans chez les personnes dont le système nerveux était délicat.—Ibid. tom. i. p. 157.

§ Strabo, lib. 14. Plinius, lib. xxix. c. 2.

municated them there, and revealed to the sick during their sleep those things which would effect their cure." "But," continues Dacier, "I should not, myself, attach much importance to the customs of a people always credulous and superstitious, unless persons of great intelligence and worthy of belief had spoken of what occurred to them in their dreams, in a manner which almost precludes doubt. Aristides attests that he was often cured by remedies which were revealed to him in his dreams. Synesius assures us, that by the same assistance he avoided great dangers. We know, too, what Socrates said of dreams."*

Among the thanksgivings which Marcus Antoninus returned to the gods, we find the following:—"I render you thanks for having given me a good father, a good mother, and good preceptors; for having made me acquainted with Apollonius, Rusticus, Maximus.—I render you thanks that I have found a wife gentle and affectionate to her husband and children; and excellent tutors for their infant education.—I render you thanks for having indicated to me different remedies during my

* Rien n'est plus commun dans les anciens que les remèdes indiqués aux malades dans leurs songes; et cela était si généralement reçu dans l'antiquité qu'on allait coucher dans les temples, croyant que les dieux se communiquaient là plus volontiers, et révélaient aux malades pendant leur sommeil les choses qui pouvaient opérer leur guérison. Mais je ne m'attacherais pas beaucoup aux coutumes des peuples toujours crédules et superstitieux, si des gens très-sages et très-dignes de foi n'avaient parlé de ce qui leur était arrivé dans leurs songes d'une manière qui ne permet presque pas d'en douter. Aristide témoigne qu'il a été très-souvent guéri par des remèdes qui lui avaient été révélés en songe. Synesius assure que par le même secours il avait évité des très-grands dangers. On sait ce que Socrate dit de ses songes.—Réflexions de l'Empereur Marc-Antonin, par Dacier. Amsterdam, 1691, p. 34.

dreams, especially for a spitting of blood and vertigo which happened at Cajeta."* The gratitude of Antoninus to Serapis for these cures induced him to have a medal struck to his honour, with an inscription, which will be found in Patin; he furthermore raised a temple, which he dedicated to him by an inscription, which will be found in Gruter:—

SERAPI DEO
M. AURELIUS ANTONINUS
PONTIFEX MAX TRIBUNIC
POT X
ÆDEM.

He also adorned a statue to the honour of the god, with the following inscription on its base:—

SERAPI SACR
IMP. CÆSAR M. AUREL
ANTONINUS AUG
PIUS FELIX COS III.
P P †

Jamblichus so distinctly describes the state of magnetic somnambulism induced by the mystic ceremonies in these temples, that the passage might be appropriately transferred into any modern treatise on animal magnetism. "This state," says he, "commences by a certain heaviness or inclination of the head forwards, the eyes close involuntarily, and we find ourselves in a state between sleeping and waking. In ordinary sleep we do not distinctly and fully remark either what occurs

* Quod per insomnia remedia mihi fuerint, indicata cum alia, tam adversus sanguinis excreationem, et capitis vertiginem, quod et Cajetæ aliquando factum est. Marc.—Antonin, lib. i. § *ultim.*

† Gruter, Inscrip. p. 85.

or where we are; but when sleep comes from divinity, we in reality sleep not. We perceive perfectly objects, and distinguish them more clearly than in our accustomed waking state; so that it is in this kind of sleep that vaticination principally takes place."* "They who are thus inspired," he continues, "do not live as other men, even in their corporeal state, because many of them do not burn when fire is applied to their bodies; the divinity that inspires them from within repels it, so that even if they are burned they are not sensible of it any more than of the punctures of iron pins, or other torments. You will see them walk in places which would be otherwise impracticable." He then proceeds to describe the different effects produced according to the different modes of inspiration. "Some of the inspired," he observes, "are tranquil, others are agitated and move the whole of the body, or only certain limbs; some dance or sing, or do quite the contrary; others again extend the body and rise themselves in the air; the sound of their voices is sometimes equal and prolonged; sometimes unequal and interrupted; sometimes strong and sometimes weak."† These effects, arising from what in that age

* Contingunt quandò somnus noster est gravedo quodam capitis, vel inclinatio, et occupatio visûs, vel medium quiddam inter vigiliam atque somnum, in somno humanis causis procedente, nos tunc non distinctè, vel plene animadvertimus aut quæ occurrunt, aut statum ibi nostrum. Quandò vero divinitus mittuntur somnia neque re verâ dormimus et assequimur, animadvertimusque *clarius*, quam consueverimus vigilantes. Ideoque in ejusmodi somnis maxime ponitur vaticinium.—Jamblic. de Myster. Lugd. 1549. fol. 52.

† Secundum prædictam inspirantium inspirationisve diversitatem inspirali alii moventur vel toto corpore vel quibusdam membris vel contra quiescunt. Item choreas, cantilenasque concilnas

was presumed to be the inspiration of the gods, manifestly correspond with the effects produced by animal magnetism as attested by the French commissioners in 1784.

But independent of the cures which were wrought in these temples by outwardly imposing religious ceremonies, we find, as I have already premised, that the curative effect of corporeal apposition, whether by the touch, the breath, or the saliva, has in all ages been recognised. When Elisha restored the child of the Shunammite to life, he lay upon it; put his mouth upon its mouth, his hands upon its hands, and he stretched himself upon the child, and the child opened its eyes. So also Naaman expected that Elisha would have stricken his hand over the place to cure the leprosy. Ælian informs us that the Psylli performed cures by placing themselves in contact with their patients, breathing into their mouths, &c.; and he adds, that those who underwent this treatment fell into a state of stupor, and were for a time deprived of their intellects. Plutarch relates of Pyrrhus, that he had in his demeanour "an air of majesty rather terrible than august," and that "he cured the swelling of the spleen by passing his right foot over the part affected, gently pressing it while the patient lay down;" he adds, that "he did not refuse to give this relief to any persons who applied

agunt, aut contra. Rursum corpus eorum vel excrescere videtur in altum vel in amplum, vel per sublima ferri. Item voces edunt vel equales, perpetuasve vel inæquales et silentio interruptas, et tum remittunt tonos tum intendunt.—Ibid. 57. We shall presently find that the diabolical arts, as they were considered, of sorcery and witchcraft, induced similar effects,—convulsions, insensibility, lucid vision, &c. The discovery of animal magnetism affords a clear solution of all these mysteries.

to him, however poor or mean they might be."* Pliny relates the same anecdote with the view of shewing that there are some persons whose bodies are endowed with medicinal properties, but he considers that it is the imagination which produces these salutary emanations.† Celsus informs us that Asclepiades could, by frictions, cure phrensy; and further states, that when these frictions were carried to a great extent, they produced a lethargic state. Cælius Aurelianus recommends manual friction for the cure of pleurisy, lethargy, and various other maladies, and describes the manner in which they should be conducted. In epilepsy, the head and forehead are to be chafed; then the hand is to be carried gently over the neck and bosom. At other times the hands and feet are to be grasped, so that we "may cure in the very act of holding the limb."‡

The emperor Vespasian cured, on the same principle, a person who was blind, and another afflicted with paralysis; the following are the facts as related by Tacitus:—"Vespasian," says he, "passed some months at Alexandria, having resolved to defer his voyage to Italy till the return of summer, when the winds blowing in a regular direction afford a safe and pleasant navigation. During his residence in that city, a number of incidents, out of the ordinary course of nature, seemed to mark him as a particular favourite of the gods. A man of mean condition, born at Alexandria, had lost his sight by a defluxion on his eyes.

* "Neque erat adeo pauper quisdam aut abjectus, cui petenti denegaret petitionem."—Plutarchus in Pyrrho.

† Plinius Hist. Nat. lib. vii.

‡ In Millingen's Curiosities of Medical Literature several of these cases are collected.

He presented himself before Vespasian, and, falling prostrate on the ground, implored the emperor to administer a cure for his blindness. He came, he said, by the admonition of Serapis, the god whom the superstition of the Egyptians holds in the highest veneration. The request was, that the emperor would, with his saliva, condescend to moisten the poor man's face and the balls of his eyes. Another, who had lost the use of his hand, inspired by the same god, begged that he would tread on the part affected. Vespasian smiled at a request so absurd and wild. The wretched objects persisted to implore his aid. He dreaded the ridicule of a vain attempt; but the importunity of the men and the crowd of flatterers, prevailed upon the prince not entirely to disregard their petition. He ordered the physicians to consider among themselves whether the blindness of the one, and the paralytic affection of the other, were within the reach of human assistance. The result of the consultation was, 'that the organs of sight were not so injured, but that by removing the film or cataract the patient might recover. As to the disabled limb, by proper applications and invigorating medicines, it was not impossible to restore it to its former tone.' The gods perhaps intended a special remedy, and chose Vespasian to be the instrument of their dispensations. If a cure took place, the glory of it would add new lustre to the name of Cæsar; if otherwise, the poor men would bear the jests and raillery of the people. Vespasian, in the tide of his affairs, began to think there was nothing so great or wonderful—nothing so improbable, or even incredible, which his good fortune would not accomplish. In the presence of a prodigious multitude, all erect with expectation, he advanced with an air of serenity, and hazarded the experiment. The paralytic hand recovered its functions; and the blind man saw the light of the sun. By living witnesses, who were actually on the spot," adds Tacitus, "both events are confirmed at this hour, when deceit and flattery can hope for no reward."* Both cases are related also by Suetonius,† who repeats that the men sought the aid of Vespasian under the advice and direction of the god Serapis; he, however, describes the paralytic limb to have been the leg, and not the arm, as mentioned by Tacitus; and also states that the cure was wrought before a public company, "*palam pro concione.*" Voltaire, who I apprehend will not be accused of a disposition to credulity, observes, "Of all miraculous cures, the best attested, the most authentic, are those of the blind person to whom Vespasian restored sight, and the paralytic person to whom he restored the use of his limb. It was not for him to attach value to prestiges which a monarch firmly enthroned does not need."‡

Spartian relates that the emperor Adrian, reduced to the last extremity by disease, came to the resolution that he would commit suicide; at which time a woman presented herself to him, who declared that she had been warned by a dream to advise him not to kill himself, for that he was about to recover; but as she did not obey this order she was struck blind.

* C. C. Tacitus Hist. lib. iv. sect. 81.

† In Vita T. Fl. Vesp. sect. 7.

‡ "De toutes les guérisons miraculeuses les plus attestées, les plus authentiques sont celles de cet aveugle à qui l'Empereur Vespasian rendit la vue, et de ce paralytique auquel il rendit l'usage de ses membres. Ce n'est pas lui qui cherche à se faire valoir par des prestiges dont un monarque affermi n'a pas besoin."

Being commanded, however, again to do the same thing, she came to Adrian to repeat it, and embraced his knees; having accomplished which dream, she washed her eyes in the lustral water in the temple, and recovered her sight.* Spartian also states that when this emperor was ill with fever, there came to him a certain blind man from Pannonia, who touched him, which being done, he who was blind recovered his sight, and the fever left Adrian.† In the same reign with Marcus Antoninus, lived Apollonius of Thyana, whose biography, written by Philostratus, supplies us with a number of other curious facts. One of the most remarkable of these was the resuscitation of a girl who was being conducted to the grave, having fallen down apparently dead during the solemnisation of her marriage. The funeral procession being incidentally met by Apollonius, he ordered it to stop, and after touching the young woman, and muttering some secret words, she recovered, began to speak, and returned to her paternal home.‡ It appears that Apollonius possessed in a high degree the magnetic power, and was himself often in a state of lucidity; read

* Ea tempestate supervenit quædam mulier quæ diceret somnio se monitam ut insinuaret Adriano ne se occideret quod esset bene valiturus, quod cum non fecisset esse cæcatam. Jussam tamen iterum Adriano eam diceret, atque genua ejus oscularetur, receptura visum si id fecisset quod cum insomnium implesset oculos recepit cum aquâ quæ in fano erat ex quo venerat oculos abluisset. —Spartianus in Adriano versus finem.

† Venit de Panroniâ quidam natus cæcus ad febrientem Adrianum, eumque contigit, quo facto et ipse oculos recepit, et Adrianum febris reliquit.—Ibid.

‡ "Ipse vero puellam attrectans atque aliquid illi secreto admurmurans ab eâ quæ videbatur morte revocavit, quæ statim etiam loqui incipiens in paternam domum rediit."—Vitâ Apollonii, lib. iv. cap. 16, fol. 175.

the thoughts of others, and perceived events passing at a distance. He was at Ephesus when the tyrant Domitian was assassinated. All at once his voice became feeble, he paused, his eyes were fixed on the ground, he advanced three or four steps forward, and exclaimed with unwonted energy, "Strike the tyrant! Strike!" And this he did, not as a person might who saw the event passing in a mirror, but as one who was present and witnessed the assassination passing directly under his eyes, and endeavoured to assist in it.* The Ephesians present were exceedingly astonished; whereupon he again exclaimed, "Have confidence, O Ephesians, for the tyrant is this day slain: but why do I say to-day? it is rather at this moment." † A few days afterwards the news arrived at Ephesus of the death of Domitian, which happened at the time and in the manner which Apollonius had described.

Hitherto, however, my attention has been restricted to the fact, that cures of a remarkable nature were wrought by the ancients by simple and direct corpo-

* "Dum hæc Romæ in regia agerentur ipse forte Ephesi in Xysti lucis meridie disputabat, et primo quidem velut repentino pavore perculsus vocem demissit deinde concise magis quam consueverat verba proferre cæpit, ut facere solent, quibus in mediâ oratione novum aliquod et incredibile occurrit. Tandem verò quasi eorum quæ dicere vellet oblitus conticuit torvis oculis in terram aspiciens, tres inde aut quatuor passus e loco ubi constiterat, progressus percute tyrrannum, percute clamabat non ut qui e speculo veritatis imaginem traheret, sed qui res ipsas ante ipsum fieri videret et auxilium affere conaretur."—Vita Apollonii, lib. viii. cap. 10.

† "Tandem confidite ô viri, inquit, Tyrannus namque occisus est hodie; quid autem dixi hodie? imò potius hoc ipso temporis momento. Ita me Pallas adjuvet. Id enim eo ipso tempore gestum est, quo me conticentem vidistis."—Ibid.

real apposition—the manipulations, *per contactum*, of the animal magnetisers—and not only from the authors I have already cited, but from a variety of other sources, we may learn that the sleep, or rather the somnambulism thereby induced, was characterised by the magnetical phenomena—if I may be allowed the phrase — of insensibility, lucidity, and prevision. Hence, in the temples of Isis and Osiris, Apollo and Æsculapius, statues, representing Sleep and Dreams as divinities, were placed in the vestibules, near which the votaries lay down to await the inspiration from the God of Health. The whole practice of medicine originally consisted in this kind of divination; and, accordingly, Jamblichus, after stating that numerous cures were performed in the temple of Æsculapius, adds, that the art of medicine itself originated in this divine sleep.* So also Philostratus, in his life of Apollonius, affirms that the art of divination rendered man essential services, the greatest of which was that of medicine;† for he adds, the sons of Æsculapius would never have acquired the art of curing, if Æsculapius, who was the son of Apollo, had not obeyed the vaticinations of his father, and thence arose the art of medicine, for by this means antidotes were discovered against poisons, and the means of making poisons themselves useful under certain circumstances. "Who can deny that this knowledge," he continues, "is referable to the art of divination? for it is not rea-

* Sic in Æsculapii templo accipiuntur somnia, quibus morbi curantur, ipsaque ars medendi somnis est comparata divinis.—Jamblic. de Mysteriis, Lugd. 1549, fol. 55.

† Jarchas de divinatione sermonem prosecutus multa bona erga homines illam efficere dixit inter quæ præcipuum est medecinæ donum.—Philostrat. de Vitâ Apollonii Lutet. 1555, lib. iii. cap. 13.

sonable to suppose that, without such foreknowledge, science would have mingled with its remedies poisons so dangerous." *

In the Orationes Sacræ of Aristides, we learn that this orator, being ill, conducted his own case entirely by the dreams which revealed themselves to him while in a somnambulic state in the temple of Æsculapius. On one occasion he was directed in his sleep to use a warm bath ;† on another, to suspend the bathing and take emetics, a practice which, on the authority of Diodorus Siculus, was very common among the Egyptians.‡ He was also, in one of his dreams, directed to be bled by his physician Theodosius, who, on visiting him, said it was necessary to obey the revelations of Æsculapius, and accordingly bled him with success.§ On another occasion he describes himself to have seen in his dream the internal condition of his body; he gives a long description of his complaint, and the treatment he adopted; and assures us that it was not until the physicians at Rome and Pergamus had failed to cure him,

* Quin etiam venenis ipsis ad multas ægritudines salubriter uti didicere. Quam perititiam qui divinationi austerre audeat nescio. Neque enim rationi consentaneum videtur sine præcognitione quadam ausos unquam aliquos fuisse tam perniciosa venena proficius medicamentis admiscere.—Philostrat. de Vitâ Apollonii Lutet. 1555, lib. iii cap. 13.

† In somnis vero visus eram in thermis esse et cum anteriùs prospicerem inferiorem ventrem male affectum videre. .... Lavi porrò vespere, ac diluculo mox doluit ventriculus procedente ad dextram dolore et usque ad inguen.—Ælii Aristides Oratoris Clarrissimi Orationes Græce et Latine interprete Gulielmo Cantero Oliva Paul Steph. 1604. Tom. i. f. 482.

‡ Morbos vel jujunio curant, vel vomitu, idque aut quotidie aut tribus diebus, aut quatuor interjectis.—Diod. Sic. lib. ii.

§ Ibid. pp. 482, 483.

and acknowledged themselves ignorant of his disease, that he had recourse to the remedies which were prescribed to him in dreams. Hence, what magnetisers term the faculty of intuition, or the power of perceiving in the somnambulic state the internal organisation—whether healthy or diseased—of the body, was occasionally observed by the ancients. It is described by Julius Scaliger with as much exactness as could be desired in any treatise professedly on animal magnetism. His words are, "When the soul is delivered by sleep from directing the grosser functions of the body, she retires within herself as into a port protected from tempests. Here she knows all that passes in the interior, that is, within the body;* she depicts as in colours, figures, and quantity, that which therein occurs;† and this is what Hippocrates, always consistent with himself, affirms, where, in his third book on Regimen, he affirms that the mind, when the eyes are closed, can discriminate those things which happen to the body."‡ We find that Galen describes the lucidity of the mind in dreams nearly in the same language as Hippocrates; he acknowledges that he was guided in the treatment of many diseases by the prescriptions which occurred to him in his dreams; and states that Hermes, of Cappadocea, in the same way, asleep in the temple named *Hephæstium*, near that of Memphis, derived the knowledge of many valuable

* Cognoscit quæ intus sunt in corpore scilicet.—Julius Cæsar Scaligeri de Insomniis commentarius in librum Hippocrates Giessaæ, 1600, p. 2.

† Ibi cum quibusdam quasi coloribus, figurisque, ac quantitatibus pingitur earum rerum quas intus nacta est.—Ibid.

‡ Quæ corpus contingunt, eadem animus cernit occulis clausis. —Hippocrates de victu, lib. iii. Geneva, 1657.

receipts. "In this manner," says Conringius, "the sick acquire the knowledge of the medicines necessary for them; and which, by their natural properties, are really adapted to their complaints. Hence arose the numerous medicaments referred to Hermes in Galen; hence also were discovered those which were indicated by Isis, which will be found not only in Galen, but in Celsus, Paul of Egina, and others." The idea that the soul, during sleep, disengages itself from the encumbrance of the body, and concentrates its energies in a higher degree than it can do in the waking state, prevailed generally among the ancient philosophers. Lucretius himself observes:—

> "Sleep then occurs when fades through every limb
> The soul's sensorial power, part by fatigue
> Wasted through ether and *concentered* part
> Deep in each vital organ."*

According to the epicurean hypothesis, there is an emanation, or effluvium, of very subtle and minute particles, thrown off from the body,† which stimulates, not only the physical organs to the due exercise of their functions, but "is capable," in the words of the late learned Mason Good, "of winding through the substance of the flesh, and of stimulating the soul itself in the interior of the animal frame, especially when in a state of sleep, in which the external sense is closed, or of deep abstraction, in which it is inattentive, and thus of presenting to the soul in its naked state, as it may be called, pictures of objects no longer in

* Lucretius de Natura Rerum, lib. iv. p. 915.

† It is impossible not to be struck with the identity between the epicurean theory and that of the magnetisers, who believe in the existence of a magnetic fluid.

existence; and hence these philosophers, though now as it appears with great incorrectness, undertook to solve many of the most difficult problems of nature, accounted for the casual appearance of spectres in the gloom of solitude and retirement, and directly unfolded to the world the 'stuff that dreams are made of.' "* How far these learned philosophers were incorrect remains, however, yet to be determined; there can be no reason why the progress of psychical science should not reveal to us the laws which determine the manifestations of the mind, whether the body be in a waking or sleeping state, as satisfactorily as physical science has made known to us many of the laws which govern the material universe. It is our ignorance alone which veils them in mystery. "The opinion of all, or most men," says Aristotle, "that dreams have some signification, procures belief as an assertion derived from experience; and that there is a divination concerning some things in dreams is not incredible, for it possesses a certain reason....... The most elegant physicians say that it is necessary to pay very great attention to dreams; and it is reasonable thus to think and believe them, since they are observed by artists, by speculative men, and by philosophers." Aristotle then proceeds to argue that dreams are not sent from divinity, but are demoniacal; "because," says he, "nature is demoniacal and not divine, of which this is an indication, that very degenerate men are prescient, and clearly see future events in their dreams." He then proceeds to discuss the cause of "persons in an ecstacy foreseeing future events," and wherefore "those who are inspired by

* Good's Book of Nature, vol. ii. p. 214. London, 1826.

prophetic fury" do the same.* Here I may observe that the matter at issue respects not the theories which ancient philosophers hazarded to explain the phenomenon, but the existence of the phenomenon itself; and hence I refer to these ancient authors in order to shew that the facts which are insisted on by the professors of animal magnetism, have in all ages been recognised.

According to Jamblichus, the soul has a two-fold life; the one which we employ when awake is "common with the body, except when by intellectual energy and the pure reasonings of the dianoctic power, we entirely depart from it;" the other, which he knows not whether to call "intellectual or divine," energises within us when asleep in such a manner as it is adapted to energise. He then continues, "if the soul connect the intellectual and divine part of herself with more excellent natures, then her phantasms will be more pure, whether concerning the gods or things essentially incorporeal, or, in short, concerning things which contribute to the truth pertaining to intelligibles. If, however, the soul refer the reasons of generated natures which she contains in herself to the gods, the causes of them, she receives from them a power and a knowledge which apprehends what was and will be, which surveys all times, beholds the production of whatever happens in time, and is allotted the province of taking care of and correcting them in an appropriate manner. And bodies indeed, which are infirm, she heals; such human affairs as are in a confused state, she aptly arranges, and frequently delivers

* Aristot. de Divinatione in Somno Lat. Francisco Vatalbo interpr. 8vo. Lugd. 1546.

the inventions of arts, the distribution of justice, and the constitutions of legal establishments. Thus, in the temple of Æsculapius diseases are healed by divine dreams; and through the order of nocturnal appearances, the medical art consists from sacred dreams. The whole army of Alexandria was saved, which otherwise would have entirely perished, in consequence of Bacchus appearing in a dream and exhibiting the solution of the deplorable calamities in which it was involved. The city of Aphutis also, when it was besieged by king Lysander, was saved by the dreams which were sent by Jupiter Ammon, in consequence of which the army rapidly marched from the city, and immediately abandoned the siege. What occasion is there," he adds, "to mention every particular of things which happen daily, and which exhibit an energy superior to that of reason. Thus much, therefore, may suffice concerning divine inspiration by sleep—what it is, how it is produced, and what benefit it affords to mankind."* In the Timæus of Plato we find that this philosopher also recognised the faculty of prevision, and even assigned it a local habitation in the liver; he observes that this organ is so constructed as to be the abode of this part of the soul, which "repairs from excess in the night, and employs prophetic energies in sleep." He adds, that "no one while endued with intellect, (in the ordinary waking state,) becomes connected with a divine or true prophecy; but this alone takes place, either when the power of prudence is fettered by sleep or suffers from mutation, through disease or a certain enthusiastic

* Jamblic. *loc. cit.* Vide Aristotle's Treatises on the Soul. Trans. by Thomas Taylor, 9 vols. 4to. London, 1808.

energy."* We learn also that Pythagoras reposed confidence in the interpretation of dreams; and Jamblichus relates that he "used means to procure men quiet sleeps with good and prophetic dreams."† It appears to me that the universal belief which prevailed among the ancients in the prophetic character of dreams, was derived from the phenomena developed during somnambulism in the temples, for, be it observed, the sleep was in these cases *artificially* induced, the votaries yielded implicit confidence in the power of the priests, and predisposed themselves to be affected by earnest prayer and resignation to the divine will. Then, again, on the part of the priests, they conducted the mystic rites—as Jamblichus has just observed—with the sincere intention and desire of producing a particular description of dreams—those which should be prophetic; so that the conditions on the part of the votaries, and those also on the part of the priests, correspond precisely with the conditions which are required on the principles of animal magnetism; that is to say, faith, or, at least, passive acquiescence, on the part of the magnetisee; and self-confidence, energy, and concentration of the desire or will on the part of the magnetiser. Nay, we have seen that Mesmer himself complicated the principle of magnetism with a variety of ceremonies and paraphernalia which were, in reality, utterly superfluous; and so was it in the ancient temples, the fumigations, sacrifices, hymns, ablutions, &c., were in reality unnecessary, and only adopted because they happened to be in accordance with the superstitions of the age.

* Plato in Timæo. Lat. Interpr. M. T. Cicerone et Chalcido, 4to. Paris, 1563.

† Jamb. *loc. cit.*

How, excepting on magnetic principles, can we interpret the following anecdote of Alexander the Great? It is related by Cicero that Ptolemy having been wounded by a poisoned arrow, was in great pain, and likely to die, when Alexander, sitting at the edge of the bed in a profound sleep, the dragon, nourished by his mother Olympia, appeared to him with a certain root in his mouth, and told him where it was to be found, which was not far from that place, and that it possessed such virtues that it would soon cure Ptolemy. On awakening, Alexander narrated his dream to his friend;—they then sought the root, which was found, and not only was Ptolemy cured, but many other soldiers who had been wounded by the same arrows.* We have already seen that magnetic somnambulists prescribe, and that successfully, for the complaints of others, and here we have precisely the same fact. A cavern is also mentioned by Strabo, to which the sick resorted for the purpose of consulting sleeping priests. We learn, too, from Cicero, that when the Lacedemonian magistrates were embarrassed in their administration, they went to sleep in the temple of Pasiphae, so called from *Pasi Phainein*, "communicative to all." To the clairvoyance and lucidity which often occur during somnambulism may be re-

* Cum Ptolemæus familiaris ejus in prælio telo venenato ictus est, eoque vulnere summo cum dolore moreretur, Alexander assidens somno est consopitus. Tum secundum quietem visus, ei dicitur draco, is quem mater Olympias alebat, radiculam ore ferre et simul dicere quo illa loci nasceretur neque is longe aberat ab eo loco; ejus autem esse vim tantam ut Ptolemæum facile sanaret. Quum Alexander experectus narasseret amicis somnium, emisisse qui illam radiculam quærerent, quâ inventâ et Ptolemæus Sanatus dicitur, *et multi milites* qui erant eodem genere teli vulnerati. —Cicero de Divin. lib. ii. No. 133.

ferred many of the visions and predictions which are recorded in ancient history. The mystery of Oracles, the prophetic raving of the Sybils, the predictions of many of the ancient philosophers, are all explicable on the same principle. That the Sybils uttered their ratiocinations in a state of magnetic *crisis* or *extase* is clear, from the account given by St. Justin, who remarks that they did not themselves understand what they said, and that as soon as the instinct which animated them was extinguished, they lost all recollection of what they had predicted.* The Sybils and Bacchides, says Aristotle, who fall into a state of enthusiasm called sacred, do so, not from disease, but only from their natural temperament.† Avicenna also repudiates the idea of their being possessed, and, fantastically enough, ascribes their power of divination to the black bile.‡ Louis Buccaferrei, of Bologna, one of the commentators on Aristotle, insists that the gift of prophecy depends entirely on the temperament, and the melancholic is the one to which he assigns the development of this faculty.§ Gaspard Pencer wrote

* Sed in ab ipso afflatus tempore sortes illa suas explebat, et evanescente instinctu ipso simul quoque dictorum memoria evanuit.—Justin, Admonitorum ad Græcos.

† Unde Sybillæ et Bachides et qui numine afflatu dicuntur cum morbo tales non fiant sed naturali temperie.—Aristot. Prob. sect. 50, prob. 1.

‡ Nos physici causam proximam naturaleno contemplantes dicimus illius causam esse atram bilem non demonem.—Avicen, lib. iv.

§ Qui habet habitum melancholicum habet per se causam prædicendi de futuris, et ideo per istum habitum prophetia erit secundum naturam et melancholicus habitus erit pro prophetâ naturaliter quia ille habitus est naturalis.—Ludov. Buccaferrei Lectiones in Aristot. lib. Venetiis, 1570, fol. 102.

a treatise professedly on divination, in which he adopts partially the theory of Buccaferrei, but insists that in many cases it is a faculty peculiar to certain individuals.* The Jews, more than any other nation of antiquity, appear to have cultivated this somnambulic power: Josephus himself was constantly in what we now term a state of crisis, that is, in the highest degree of somnambulism, in which prevision is developed. On the authorities of Suetonius, Dio, and all Roman historians, he foretold the succession of Vespasian and Titus to the Roman empire, and this he did in the days of Nero, before Galba, Otho, or Vitellius, were thought of to succeed him. His faith in the lucidity of such visions was manifested on the occasion of his being in imminent danger after the fall of Jotapata. When the city had been taken, he sought refuge from the enemy in a deep pit, adjoining a large cavern, which could not be seen above ground. Here he met with forty persons of eminence, who had also, in a similar manner, concealed themselves; but on the third day he was betrayed by a woman, upon which Vespasian sent two tribunes to him, and ordered them to give Josephus their right hands as a security for his life, and to exhort him to come up. He was, however, afraid to comply with the request, lest he should be punished; and Vespasian then sent Nicanor, a personal acquaintance of Josephus, to persuade him that no such perfidy would be committed. Thereupon the enraged soldiery threatened to set fire to the cavern, which the tribune would not permit, being desirous that Josephus should be taken alive. "And now," I quote the history itself, "as Nicanor lay hard

* "Sunt qui prœvidunt futura et intelligunt," &c.—Ludov. Buccaferrei Lectiones in Aristot. lib. Venetus, 1570, fol. 102.

at Josephus to comply, and he understood how the multitude of enemies threatened him, he called to mind the dreams which he had dreamed in the night-time, whereby God had signified to him beforehand both the future calamities of the Jews, and the events which concerned the Roman emperors. Now Josephus was well able to give shrewd interpretations of such dreams as have been ambiguously delivered by God. Moreover he was not unacquainted with the prophecies contained in the sacred books, as being a priest himself and of the posterity of priests. And just then was he in an ecstacy; and, setting before him the tremendous images of the dreams he had lately had, he put up a secret prayer to God, and said, 'Since it pleaseth thee, who hast created the Jewish nation, to depress the same; and since all their good fortune is gone over to the Romans; and since thou hast made choice of this soul of mine to foretel what is to come to pass hereafter, I willingly give them my hands, and am content to live; and I protest openly, that I do not go over to the Romans as a deserter from the Jews, but as a minister from thee.' When he had said this, he complied with Nicanor's invitation."* From what follows, it appears that Vespasian himself at last gave credit to his predictions. Nay, when Josephus was led by Nicanor before Vespasian, we almost recognise a direct magnetical action on those by whom he was immediately surrounded: they who were remote from him, cried out that he should be put to death; but they who were near his person, relented, and were subdued.†

* Josephus de Bello Judaico, lib. iii. cap. 12. Vide etiam Dion Cassius, lib. lxvi. et Suetonius in vita Vesp.

† Ibid. lib. iii. cap. 14.

The prevision of the dying was well known to the ancients. Possidonius cites the example of a Rhodian who on his death-bed predicted to six companions who were in attendance the order in which they would severally die.* Quintus adds the case of the Indian Calamus. This philosopher, at eighty-three years of age, being very ill, and in great pain, determined to commit suicide. On ascending the pile, he exclaimed, "How delightful is it to depart this life! When my body is reduced to ashes, like that of Hercules, my spirit will emerge into light." Then Alexander asked him if he wished to say anything, to which he replied, "Yes—speedily I shall see you!" which happened accordingly, for in a few days afterwards Alexander died at Babylon.† So also Apollonius foretold the day and hour when Cocceius would die, and the time and manner also of the death of Titus Vespasian. Melancthon reports, that a woman predicted that Dioclesian would become Emperor of Rome. Urban Grandier foretold the death of Father Lanctantius, and his prediction was verified. Indeed, a host of facts of the same description might here be adduced, which are not, in my judgment, mere happy guesses, or accidental coincidences, but revealed by

* Divinare autem morientes etiam illo exemplo confirmat Possidonius, quo affert, Rhodium quemdam morientem sex equales nominasse et dixisse qui primus eorum qui secundus, qui deinceps moriturus esset.—Ibid. No. 64.

† Ad mortem proficiscens, Calamus Indus quùm ascenderet in rogum ardentem, Oh Pulcrum discessum, inquit, e vitâ quùm ut Herculi contigit mortali corpore cremato in lucem animus excesserit. Quùmque Alexander eum rogaret si quid vellet ut diceret:—Optimè, inquit, propediem te videbo. Quod ita contigit. Nam Babylone paucis post diebus Alexander est mortuus. Ibid. No. 47.

the same faculty of lucidity which enabled the epileptic to predict to the French commissioner, in 1835, the day, the hour—nay, the very moment—when his next paroxysm would occur.

The somnambulic faculty of clairvoyance, or the power of seeing events passing at a distance, affords a solution of the mystery of what, in the north of Scotland, is called *second sight,* and many occurrences of this description are on record. We have seen that Apollonius, who was habitually *crisiaque* when at Ephesus, saw the assassination of Domitian. So also we learn from Tacitus, that, when Vespasian, during his sojourn at Alexandria, went to consult Serapis, to know whether he should be emperor or not, Basilides, the Egyptian, appeared by his side; albeit, he was at that moment eighty miles from Alexandria.*

St. Augustin relates a still more remarkable example:—A man, well educated, who occupied himself particularly in studying the philosophy of Plato, assures us, that one night, in his own house, before he fell asleep, a philosopher, well known to him, appeared before him, and explained a number of difficulties in platonism which he had before in vain solicited him to propound. When the narrator afterwards met the philosopher, he asked him how it happened that he explained to him in his house that which he had refused to do in his own; to which the philosopher replied, "I did nothing of the sort; I only dreamed that I did so." By this means, adds St. Augustin, the one who was awake really saw the fantastic image which appeared to the other in his dreams; this, to us, would be unworthy of credit, but those to whom

* Tacitus, Histor. lib. iv. No. 82.

we refer we do not consider capable of deceiving us.* This vision, related by St. Augustin, may recall to the recollection of those acquainted with the history of animal magnetism a case which is reported by Deleuze, in the 13th number of the Bibliothèque du Magnétisme. It occurred at Stutgardt, and is attested by Dr. Meier, of Carlsruhe, and Dr. Klein, counsellor to his Majesty the King of Wirtemberg. "Mademoiselle Müller, without quitting her bed," says Deleuze, "appeared to a friend of hers in a house at a distance, and cured her of the tooth-ache. She affirmed that it was her spiritual being which, conducted by the soul of her mother, rendered this visit."† All magnetisers believe that magnetised persons can tacitly communicate with, and know intimately each other's most secret thoughts. It is even affirmed, that in this state of exaltation, the mind of the person magnetised is so much divested of all gross and sensual qualities, that it recoils from the slightest impurity in the ideas of any person with whom it may be in magnetical connexion. "Such persons," says Loewe, "express this without reserve, and add, that those individuals who would spare them the pain of violent spasms, must relinquish their impure ideas, or

* Indicavit et alius se domi suæ per noctem antequam requiesceret, vidisse venientem ad se quemdam philosophum, sibi notissimum, sibique exposuisse nonnulla Platonica quæ antea rogatus exponere noluisset. Et cum ab eodem philosopho quæsitus fuisset cur in domo ejus fecerit quod in domo sua petenti negaverat "*non feci* inquit *sed me fecisse somniavi.*" Ac per hoc alteri per imaginem phantasticam exhibitum est *vigilanti* quod *alter* vidit in somnis. Hæc ad nos quibuscumque qualibus credere putaremus indignum sed eis referentibus pervenerunt quos nobis non existimaremus fuisse menitos.—De Civit Dei, lib. xviii. c. 18.

† Bibliothèque du Magnétisme, No. xiii. p. 43.

leave the room."* Hence it is that purity of intention on the part of the magnetiser is insisted on as one of the elementary conditions under which the operation should be performed.

In the records of history, ancient and modern, examples of clairvoyance, or what Ennemoser and Wolfahrt describe as universal lucidity, are very numerous. Aulugelle relates, that during the war between Cæsar and Pompey in Thessaly, a priest, well known for the sanctity of his life, suddenly cried aloud in the town of Padua, that he saw the battle terminating; that the one party fled, the other pursued; that he heard the cries of the wounded and dying; and then, as if himself taking part in the combat, he exclaimed, "Cæsar is conqueror." He was looked on as insane; but it afterwards appeared that the day and hour of the battle, and alternations of victory, happened exactly as he described.† Eunapius, who lived under the emperor Julian, in his life of Eclesius, relates an incident of the same description, which occurred to the celebrated Sosipatra. One day, while discoursing on the nature of the soul, she suddenly paused, and, after a moment's silence, exclaimed, "What is this? My friend, Philometer, has fallen from his car, and is in danger of having his limbs broken; but, no, he has only grazed his elbow and hands—the danger is now over—he is put safely into a chaise;"—all which happened at the time, and in the manner she described.‡

* "I was myself present," adds Loewe, "when a magnetised person said to the magnetiser, 'No, Doctor, you have impure ideas; I beg you will leave them; you give me much pain.'"—*Op. cit.* Append. p. 129.

† Aulugelle, liv. xv. ch. 18.

‡ Eunapius in Eclesio, p. 59. Genevâ, 8vo. 1616.

Alexander, of Alexander, relates that a young man, named Marius, his pupil, one night sleeping in bed with him, suddenly awoke in tears, and cried out, that he had just seen his mother breathe her last, and the preparations for her funeral. Alexander noted the day and hour. A few days afterwards a messenger arrived, who brought intelligence of the event, which happened at the precise time the young man had intimated.* Bodin relates the case of a young man named Jacques, a priest at Perouse, who one day, while celebrating mass, instead of saying, when he turned before the people, "Brethren, let us pray," exclaimed, "Pray! for the army of the church is in extreme danger;"—at which time the army, twenty-five leagues from Perouse, was actually defeated.† It is also related that St. Ambrose, when he should have been celebrating mass, fell asleep by the altar, and remained sleeping three hours: on being awakened, he said to those around him, "Be not troubled; it was important that I should sleep, for God hath revealed to me a great miracle; know that my brother St. Martin is dead. I, myself, assisted at his funeral obsequies." The people were astonished; and having noted the day and hour, it was found that at this precise time the bishop was being interred.‡ A remarkable example of this lucid vision will be found in the Memoirs of the Queen of Navarre; it is quoted by Deleuze, in his Histoire Critique du Magnétisme. The following are the particulars:—The queen, my mother, was at Metz dangerously ill with fever. She

* Alexander ab Alexandro Genialium dierum, lib. i. c. 2.

† Bodin dans sa Réfutation de Jean Wier à la suite de sa Démonomanie, édit de Niort. 1616. 8vo. p. 547.

‡ Gregor. Turen. de Miraculis. Paris, 1640. 12mo. lib. i. cap. 5.

was delirious; and around her bed were my brother and sister, my brother of Loraine, several ministers and ladies, and princesses, all of whom gave up all hope, but did not leave her. She cried in her reveries, as if she witnessed the battle of Jarnac. "See how they fly—my son has the victory. Oh, my God! help my son—he is down! See you not in the turmoil the Prince de Condé dead?" All present thought she raved; but the night afterwards M. de Lopez arrived with the intelligence. "I knew it well," said she; "did I not see it yesterday?" From this we learned that what she had uttered proceeded not from the delirium of fever, but from the warning which God gives to illustrious personages.*

In the German Annals of Medicine, Schelling relates the following case, which occurred under his own observation:—"In a crisis of clairvoyance, Miss M., having previously been quite cheerful, began, all at once, to assume an appearance of anxiety and sorrow, and at last fell a-weeping. When I asked her what was the matter, she answered, that she had just then become aware that a death had recently taken place in the family, at the distance of more than one hundred and fifty leagues. I endeavoured to dissuade her from entertaining such thoughts, but in vain; she insisted that she was quite certain of the fact, and continued to weep. Wishing to ascertain how she had come by this intelligence, she said she herself did not well know, but that she had at once become quite certain of it. I asked her whether she could name the person who had died: she said she could not at that time, but should be able to do so in a future

* Mémoires de Marguerite de Valois, Reine de Navarre, p. 84.

crisis. She added, that the letter containing the intelligence was then upon its way. She conjured me to say nothing about this presentiment after the crisis, otherwise it would give her mortal anxiety.

"It is well known," says the Professor, "that somnambulists, when they awake out of the magnetic sleep, have not the slightest recollection of what may have taken place in it. When she awoke out of her sleep, Miss M. was as cheerful as ever, and had not the most distant idea of her vision. The expression of pain she exhibited during the crisis, which seemed to proceed so entirely from an internal conviction of the reality of the fact, and the obstinacy with which she adhered to her assertion of its truth, induced me to give her credit. I mentioned the case to Professor Schmidt, in order that he might be a witness to the fact. I awaited with great anxiety the hour when I could again set my somnambulist asleep, in order to ascertain whether she would again have the same vision. For a considerable period during the crisis, nothing of the kind appeared. She was, as usual, quite cheerful, and spoke a great deal, until, all at once, marks of sorrow were exhibited in her countenance. She turned away her face, and hid it on her arm, which she had placed on the arm of her chair, and wept in silence. At length, I asked her what ailed her. 'The same as yesterday,' she answered; 'a death has taken place in our family—I know it for certain.' She thought it fortunate for her that she knew nothing of this when awake, because it would occasion her so much grief. I again asked her whether she did not also know the individual who was dead; and she repeated that she would be able to tell me in a future crisis, provided I put the question to her. In the following crisis, as soon as she was set asleep, she again began to weep. She requested me to use every means of diverting her attention from this circumstance during her sleep, and I endeavoured to do so by introducing other subjects of conversation; yet she frequently reverted to it. Had I foreseen the circumstances which subsequently made it impossible for me to place her more frequently in a state of crisis, I should, upon the last occasion, instead of diverting her thoughts from the subject in question, rather have endeavoured to ascertain whether she could give any further particulars of the event. But I neglected the opportunity, and reserved my questions for future crises, which could no longer take place. Four or five days after the last sitting, upon entering her apartment, I found Miss M. much downcast, with appearances on her countenance indicating that she had been weeping. On inquiring into the reason of this, she pointed to a letter which lay upon the table, and said it contained intelligence of the death of a near relative and particular friend. I asked her whether she had received any previous accounts of the indisposition of this individual. She answered, 'No—none at all; the intelligence came upon me quite unexpectedly.' Nor was she at all aware of any presentiment she had of the event."

Dr. Arndt, the eminent German physician, relates, that being one day seated near the bed of one of his somnambulists, on a sudden she became agitated, uttered sighs, and, as if tormented by some vision, exclaimed, "O heavens! my father! he is dying!" A few moments afterwards, she awoke, seemed quite cheerful, and recollected nothing of the anxiety she had so recently manifested. She again relapsed twice into the same state of magnetic sleep, and each time she was tormented by the same vision. Being asked what had happened to her father, she answered, "He is bathed in blood—he is dying." Soon afterwards she awoke, became composed, and the scene finished. Some weeks afterwards, Dr. Arndt found this lady pensive and sorrowful. She had just received from her father, who was at a distance of some hundred miles—an account of a serious accident which had befallen him. In ascending the stair of his cellar, the door had fallen upon his breast—a considerable hæmorrhage ensued, and the physicians despaired of his life. Dr. Arndt, who had marked the precise time of the preceding scene of the somnambulism of this lady, found that it was exactly on the day, and at the hour, when the accident happened to her father. "This," observes the Doctor, "could not have been the mere effect of chance; and, assuredly, there was no concert nor deceit on the part of the observer."

All those who have studied the philosophy of dreams are aware that, even in ordinary sleep, our dreams not unfrequently assume a high degree of lucidity. Almost every individual will remember instances of dreams being verified in a wonderful manner. Dr. Abercrombie, in his work on the Intellectual Powers, relates several such dreams, and candidly acknowledges that he is unable to account for them. "Cases of dreams," says he, "are on record, which do not admit of explanation on any principle which we are able to trace. Many of these histories, there is every reason to believe, derive their marvellous character from embellishment and exaggeration; and in some instances which have been related to me, I have found this to be the case after a little investigation. Others, however, do not admit of this explanation, and we are compelled to receive them as facts, which we can in no degree account for. Of this kind I shall only add the following example, and I shall do so without any attempt at explanation, and without any other comment than that its accuracy may be relied on in all its particulars.

"Two ladies, sisters, had been in attendance for several days upon their brother, who was ill of a common sore throat, severe and protracted, but not considered as attended with danger. At the same time one of them had borrowed a watch from a female friend, in consequence of her own being under repair; this watch was one to which particular value was attached, on account of family associations; and some anxiety was expressed that it might not meet with any injury. The sisters were sleeping together in a room communicating with that of their brother, when the elder of them awoke in a state of great agitation, and having roused the other, told her that she had had a frightful dream. "I dreamed," she said, "that Mary's watch stopped, and that when I told you of the circumstance, you replied, 'Much worse than that has happened, for ——'s breath has stopped also;'" naming their brother, who was ill. To quiet her agitation, the younger sister immediately got up, and found the brother sleeping quietly, and the watch which had been carefully put by in a drawer, going correctly. The following night, the very same dream occurred, followed by similar agitation, which was again composed in the same manner—the brother being again found in a quiet sleep, and the watch going well. On the following morning, soon after the family had breakfasted, one of the sisters was sitting by her brother, while the other was writing a note in the adjoining room. When her note was ready for being sealed, she was proceeding for this purpose for the watch alluded to, which had been put by in her writing desk; she was astonished to find it had stopped. At the same instant she heard a scream of intense distress from her sister, in the other room,—their brother, who had still been considered as going on favourably, had been seized with a sudden fit of suffocation, and had just breathed his last."* On the principles of animal magnetism, facts of this description are explicable, for as I have already stated, natural sleep is the initiatory condition of somnambulism, and the same phenomena often develop themselves spontaneously, which occur during the magnetic state.

We have seen that the cures which were wrought in the Egyptian, Greek, and Roman temples, were, perhaps unconsciously on the part of the priests, effected upon what we *now* recognise to be magnetic principles; but more than this, it would appear as if all nations—the most unlettered and barbarous—instinctively adopted the magnetic method of treating diseases. Their charms, their incantations, their ceremonies, however whimsical, involved a principle of which they may have been ignorant; but as no effect can be produced without a cause, so, unless an active principle had been developed, all such rites must have proved ineffective. Among the ancient Oriental nations, the treatment and cure of diseases by the application of the hands is universally attested. Jarchas informs us, that he saw almost every description of disease cured in this way by the Indian sages. Some of the cases he has

* Abercrombie, *Op. Cit.* pp. 280, 281.

detailed are interesting. "A stout man," says he, "thirty years of age, a great hunter of lions, was severely wounded in the chase by one of these furious animals, who turned round on him, and severely lacerated, and even dislocated, the thigh. And how was he treated? They reduced the limb, and relied for the cure entirely on manual frictions, under which he perfectly recovered." Jarchas also cites the case of a person whose sight was restored by the same process; and that of a man whose hand having been paralysed recovered its use in the same manner.* Jobson informs us, that the Mandingos, or Æthiopians, on the banks of the river Gambia, have no recourse to herbs, or any description of medicines, but trust entirely to charms, in the treatment of every disease, no matter whether surgical or medical. "The gregories," he observes, "bee things of great esteeme amongst them; for the most part they are made of leather, of several fashions, wounderous neatly; they are hollow, and within them is placed and sewed up close, certaine writings or spels, which they receive from their Marybuckes (or priests), whereof they conceive such religious respect that they doe confidently beleeve no hurt can betide them whilst these gregories are about them. .... If any of them be possessed of any malady, or have any swelling or sore upon them, the remedy they have is only by placing one of these blessed gregories where the griefe lies, which they conceite will helpe them; and, for aught I can perceeve, this is all the physicke they have amongst them. .... To countenance his state he hath many times two of his wives sitting by him, supporting his

* Vita Apollonii Zenobio Acciolo interprete Lutet, 1555, lib. iii. c. 12.

body, and laying their hands upon his naked skin, above the waist, stroking and gently pulling the same, which he seems to receive with content."* We learn too, from one of the latest, and perhaps best works which has appeared in this country on China and the Chinese, by Charles Toogood Downing—that animal magnetism, as a remedial agent, is still had recourse to by this curious people.†

In the heathen temples, and even in the catholic church, several of the forms observed during the solemnisation of religious rites appear to be derived from magnetism as practised in the early ages. "We find in the Mosaic history," says the Reverend James Blunt, "that the rod was peculiarly appropriated to purposes of magic; it was used by the sorcerers and enchanters of Pharaoh. .... It could communicate a blessing as well as a curse; it could remove as well as inflict a calamity." He then proceeds to notice another distinctly magnetical action. "Thus am I led to mention another ancient charm to which credit is yet given in Italy. Nothing is more common than to see the Romans approach the confessionals and receive a touch on the head from a wand, which the father confessor holds in his hand, and which is supposed to confer some benefit on the parties. I believe that the priests of Isis imparted a blessing in

* The Golden Trade; or, a Discovery of the River Gambia and the Golden Trade of the Æthiopians. By Richard Jobson. London, 1623. p. 51.

† "We are informed that the subjects of his imperial majesty have resorted to the science of Mesmer for the cure of diseases, and that animal magnetism is frequently practised as a remedial agent."—The Fanqui in China, 3 vols. 8vo. By Charles Toogood Downing, Esq., M.R.C.S., &c. London, 1838. Vol. ii. p. 165.

nearly the same manner; and that such was the legitimate method of exorcism may be argued from the way in which the companions of Ulysses describe themselves to have been restored to their natural forms:—

> 'Sprinkled with better juice, her wand reversed
> Above our crowns, and charms with charms dispersed,
> The more she sings we grow the more upright.'"*

The learned author notices particularly the mysterious virtues ascribed to the human saliva. He refers to the authorities of Pliny and Varro, and then adds, "It is curious that to this day a set of men exist in Sicily, who profess to heal the wounds of animals by their spittle. They frequent the neighbourhood of Syracuse, and, as I was informed, annually assemble in great numbers at Palazzuolo, a place at some distance from that city, on the festival of St. Paul, their patron Saint. .... It is remarkable that in administering the rite of baptism, the priest, among other ceremonies, moistens a napkin with his own saliva, and then touches with it the eyes and nose of the child, accompanying the action by the word, *Ephphatha*. It was with a similar rite that Roman infants received their names on the Dies Lustricus."† It is reasonably

* Spargimur innocuæ succis melioribus herbæ
Percutimurque caput conversæ verbere virgæ
Ovid. Met. xiv. 300.

It is worthy of observation that the attitudes of the priests and their votaries, in the temples of Isis and Æsculapius, as copied by Patin, Gruter, and others, correspond with the position of the magnetiser operating in the form now prescribed upon the patient.

† Vestiges of Ancient Manners and Customs discoverable in Modern Italy and Sicily. By the Rev. John James Blunt, Fellow of St. John's College, Cambridge. London, 1823. pp. 167. 171.

conjectured by M. Colquhoun, that on the fall of the Roman empire, when literature, science, and the arts, took refuge in monasteries, the secrets and practices of the ancient Egyptians, Greeks, and Romans, which had previously been confined to the temples, passed also into the monasteries, and the magnetic mode of treatment was thus preserved. Hence, St. Jerome complains that the *leger-de-main* practices which had been adopted in the heathen temples, were passed into, and desecrated the primitive church. So also Thiers informs us that Protogenes, priest of Edessa, cured the children, his pupils, by prayer and the touch of his hand; and that the monk, John, had received from God the gift of curing the gout, and replacing dislocated limbs. The monk Benjamin cured all kinds of diseases by the touch of his hand, and anointing with holy oil. Petrus Thyraeus, the Jesuit, in his work on Dæmoniacs, refers to a number of cures performed by ecclesiastics by the imposition of the hands and other means analogous to the magnetic.*

Accordingly, the early history of animal magnetism throws considerable light on many of the mystic rites and ceremonies of the ancients; it unseals the door of the heathen temple, and strips the priest before his altar of all his supernatural attributes; it explains a variety of extraordinary facts which the historians of all ages have recorded, and connects them with each other as effects which are identical and dependent on the operation of the same cause. It affords, too, as I shall now proceed to shew, a satisfactory explanation of the curious facts connected with sorcery and

* Vide Isis Revelata, vol. i. p. 192.

witchcraft, which, during the middle ages, excited the apprehension of the vulgar, imposed on the credulity of the learned, and led to those horrible persecutions which prevailed throughout Europe. This melancholy chapter in the history of mankind is replete with details which can only be comprehended on magnetic principles.

CHAPTER IX.

SORCERY AND WITCHCRAFT.

In studying the origin of any science, when, in retracing the steps of human knowledge, we arrive at the superstitious notions which were entertained by our ancestors, it is incumbent on us, instead of turning away from them with ridicule and contempt, to investigate their nature and the manner in which they originated; for it may be laid down as an axiom that every popular belief, however extravagant it may become, springs originally from some fixed principles in nature. These for ages may be misinterpreted and obscured by all manner of vague idealisms; nevertheless, the elements of truth remain unobliterated; the ore may be yet separated from the dross, and the principle which worked in darkness fairly and clearly brought to light. Hence the history of superstition itself is an instructive record; nay, it is wonderful what germs even of divine truth will often be found scattered through masses of apparent fiction.

In looking over the trials of those unfortunate persons who were condemned for sorcery and witchcraft, it is impossible to escape the conviction that the sorcerers or wizards exercised a truly magnetic influence over their credulous disciples. Far be it from me to strain the interpretation of the facts in favour of magnetism; I am myself surprised to find the evidence so clear, so full, and so conclusive.* The persons who were so affected will be found to have exhibited all the phenomena of magnetic somnambulism, complete physical insensibility, convulsions, sometimes partial, sometimes general, sleep-talking, clairvoyance, prevision, and it is certainly not at all surprising that effects so extraordinary should in this age of darkness have been ascribed to demoniacal possession. Furthermore, on the part of those who professed these supposed satanic arts, it will appear manifest, that in the exercise of their skill, they discovered the development of a power within themselves to affect others, of which they were originally unconscious, and which confirmed them in the idea of their being really endowed with supernatural power, so that many who commenced their career as conjuring impostors, eventually died the victims of a faith, which neither the terrors of the scaffold nor the stake could induce them to repudiate. In confirmation of these views I now proceed to adduce cases, many of which indeed require no comment.

In the year 1599, a girl named Martha Brossier, of Romorantin, in Berry, was reputed to be possessed, and excited a considerable sensation in Paris. At the suggestion of the right reverend the bishop of Paris, the king ordered a committee, composed of the most eminent physicians, to examine and report on her case. The physicians appointed were Marescot, Ellain, Haulin, Riolan, and Duet; and their report, which is exceedingly curious, will be found translated

* To the researches of the Translator I have to ackowledge myself indebted for the very curious and interesting facts contained in this chapter.

into English by Abraham Hartwel.* The dedication to his majesty—I quote from this English translation—proceeds thus:—" Sir, by the commandmant of your majestie wee have set down briefely and truly that which we have found in our visiting of Martha Brossier. .... We present the same unto your majestie without any art, without any painted show, without flourish, but with a naked simplicitie, the faithful companion of truth, which you have desired from us in this matter, and which in every other matter you have always loved and curiously sought." The report then continues, " We, the undersigned Doctors Regents in the facultie of physicke, in the Universitie of Paris, touching the matter of Martha Brossier, a maide of the age of two-and-twenty yeres, or thereabouts, born at Romorantin, in Berry, who was brought unto us in the chappel of my Lord of Saint Geneufue .... and who we saw sometimes in constitution, countenance, and speech, as a person sounde of bodie and minde, ... do say in our consciences, and certify that which followeth; that all which is before set down, [referring to the character of her fits] must be referred to one of these three causes, sicknesse, counterfeiting, or diabolicall possession."

" For the opinion that it proceedeth from sicknesse, we are clerely excluded from that, for the agitations and motions we observed therein, doe retain nothing of the nature of sicknesse, nay, not of those diseases whereunto of the first sight they might have resembled; it being neither an epilepsie or falling sicknesse, which always supposes the loss of sense and

* A True Discourse upon the Matter of Martha Brossier, of Romorantin, translated out of French into English, by Abraham Hartwell. London, imprinted for John Wolfe, 1599.

judgment, nor the passion which we call hysterica, or suffocation of the matrix, nor any of the foure motions proceeding from diseases,—that is to say, shivering, trembling, panting, and convulsion, or, indeede if there doe appeare any convulsion, and that a man will so call the turning up of her eyes, the gnashing of her teeth, the writhing of her chaps, which are almost ordinarie with this maide while she is in her fittes; the confidence which the priest hath when he openeth her mouth, and holdeth it open with his finger within it, testifying sufficiently that they doe not proceede from, nor are caused by, any disease, considering that in diseases he that hath a convulsion is not master of that part or member wherein it is, having neither any power of election or command over it, and particularly which is in the convulsion of the jawes, which is most violent of all the rest, the finger of the priest should bee no more respected nor spared than the finger of any other man. Moreover, diseases, and the motions also of diseases (especially those that are violent), leave the body feeble, the visage pale, the breath panting. This maide at the end of her fittes was found to be as little moved and changed in pulse, colour, countenance, and breath, as ever she was before; yea, which is the more to be noted, as little at the end of her exorcisme as at the beginning, at evening as in the morning—at the last day as at the first."

"Touching the point of counterfeiting, the insensibilitie of her bodie during her extasies and furies tried by the deepe prickings of long pinnes, which were thrust into divers parts of her hands and of her necke, and afterwards plucked out againe without any show that ever she made of feeling the same, either in the putting in of them, or the taking out of them, a griefe which, without majicke and without speach, could not, in our opinion, be indured without any countenance or show thereof, neither by the constancie of the most courageous, nor by the stoutnesse of the most wicked, nor by the stronge conceit of the most criminall malefactors, took from us almost the suspicion of it, but much more persuaded us from that opinion, the thin and slender foam that in her mad fits we saw issue out of her mouth, which she had no means to be able to counterfeit. And yet more than all this, the very consideration before mentioned of the little or no change at all that was seene in her person after all these most sharpe and very long pangs (a thing which nobody in the world did ever trie in their most moderate exercises), we are driven even till this houre, by all the lawes of discourse and knowledge, yea, and almost forced to beleeve, that this maide is a demoniacke, and the devill dwelling in her is the author of all these effects. .... If wee had seen that which my lord of St. Geneuefue, and many others, doe report that this maide was lifted up into the ayre more than four foote above five or six strong persons that held her, it would have been an argument to us of an extraordinarie power over and beyond the common nature and condition of man. But not being presente at that wonder, we doe give a testimonie of our knowledge, which is as much, or rather more admirable than that force and power was,—viz., that being demanded, and in her exorcising commanded—my lord of Paris furnishing the priest with questions and interrogatories—this maide divers and sundrie times, by many persons of qualitie and worthie of credit, was seene and heard to obey and answere to purpose, not only in the Latin tongue (wherein it had not been impertinent, peradventure, to have suspected some collusion), but also in Greeke and in English, and that upon the sudden. She did (we say once againe) understande the Greeke and English languages, wherein we beleeve, as it is very likely that she was never studied, so that there was no collusion used with her, neither could she invent or imagine the interpretations thereof. It resteth therefore even in the judgment of Aristotle in the like case that they were inspired unto her." The report then concludes with this solemn declaration;—"By reason whereof, and considering also (under correction) that Saint Luke, who was both a physician and evangelist, describing the persons out of whose bodies our Lord and his apostles did drive the devils left unto us, none other nor any greater signes than those which wee thinke wee have seene in this case; wee are the more induced, and almost confirmed to beleeve and to conclude as before, taking God for a witnesse of our consciences in the matter.—Made at Paris, this 3rd April, 1599."

On this report I have only to observe that the physicians Marescot, Ellain, Haulin, Riolan, and Duet, were all men of scientific attainments, and unimpeachable moral integrity;—the same facts were also witnessed and attested by the bishop of Paris, the abbot of Geneuefue, and other competent observers.

In the year 1564 the case is reported of a girl named Anne Myllner, and some curious facts connected with it are attested by Sir William Calverly, Sir William Sneyd, Lady Calverly, and other persons of distinction who lived at Chester. The description of the paroxysm is extremely graphic. "We went (says the report, which is signed by the above parties) at about two of the clocke, in the afternoone of the same 16th day of February, and there found the mayden in her traunce, after her accustomed manner, lying in a bed within the haule, her eyes half shut, half open, looking as she had been agast, never moving either eye or eyelid, her teeth something open, with her tongue doubling betweene, her face somewhat red, her head as heavy as leade to lift at, there she lay stil as a stone, and feeling her pulse it beat in as good measure as if she had been in perfite health." The report then describes her becoming violently convulsed:—" She lifted herself up in her bed bending backwards in such order that almost her head and fete met, falling down on the one side then on the other." A person of the name of Lane, who was reputed to possess great power over demoniacs, is then called in, who first, as the report expresses it, "*willed*" that she should speak, and then "*willed*" that she should rise and dress herself, all which she did to the astonishment of the bystanders, and a certificate to that effect was signed by all present on March 8, 1564.*

The violence of the convulsions is particularly noticed in all such cases;—in that of Anne Styles, bewitched by Anne Bodenham, who was executed at Salisbury in 1653, the narrative states that "she was so strong in her fits that six men or more could not hold her;" but while suffering under "most grievous hurryings and tortures of the body," the witch being only brought into the room "she fell asleep, and slept for three hours, so fast that when they would have

* The copy of a letter describing the wonderful worke of God in delivering a mayden, within the city of Chester, from a horrible kinde of tormente and sicknesse, 16th February, anno 1564. Imprinted at London for John Judely, dwelling in Little Britayne Streete, beyonde Aldersgate, 23rd March, 1564.

awakened her they could not."* The insensibility of the body in this state, we are informed by Mather, led to a cruel test for demoniacal possession. "There was a notorious witchfinder," he observes, "in Scotland, who undertook by a pin to make an infallible discovery of suspected persons whether they were witches or not. If, when the pin was run an inch or two into the body of the accused party, no blood appeared, nor any sense of pain, he declared them to be witches, by means of which 300 persons were condemned for witchcraft in that country."†

In a treatise entitled Daimonomagia, 1665, the effects of witchcraft are described as a disease, and the little pamphlet sets out with this definition:—"A disease of witchcraft is a sickness that arises from strange and preternatural causes, and from diabolical power in the use of strange and ridiculous *ceremonies* by witches or necromancers, afflicting with strange and unaccustomed symptoms, and commonly preternaturally violent, very seldom or not at all curable by natural remedies." We have then the signs diagnostical, among which are convulsions, insensibility, preternatural knowledge of languages. Afterwards we have the causes—"Biernannus and Wierius find that aspect and contact do not bewitch; but witches sometimes try to bewitch one and cannot, and yet bewitch another of the same family." Lastly, touching the cure, we are instructed to make the witch, besides prayers, charms, and certain

* An Antidote against Atheism, or, An Appeal to the Natural Faculties of the Mind of Man. By Henry More, Fellow of Christ's College. 1655.

† Cases of Conscience concerning Evil Spirits personating Men. By Increase Mather. Printed at Boston; and reprinted in London for John Dutton, at the Raven, in the Poultry, 1693.

herbs, use the ceremonies by which she knows how to cure the sick.* In the fact that the sorceress could affect one member of a family and not another, we recognise that difference in the magnetic susceptibility of different persons which modifies so much the facility with which the effects are produced.

Among other unfortunate women who were arraigned for practising witchcraft, we find the case of one Jane Brookes, who was tried, condemned, and executed at Charde, March, 1658. She was indicted for bewitching Richard, the son of Henry Jones, of Shepton Mallet, and numerous persons, including physicians and clergymen, witnessed his paroxysms. In this case we shall find that the direct personal influence of the woman exercising what was imagined to be a demoniacal power was clearly proved. "The boy," I quote Glanville's report, "fell into his fits on the sight of Jane Brookes, and lay in a man's arms like a dead person; the woman was then willed to lay on her hand, which she did, and he thereupon started and sprung out in a very unusual manner. One of the justices, to prevent all possibilities of *legerdemain*, caused Gibson and the rest to stand off from the boy, and then that justice himself held him; the youth being blindfolded, the justice called as if Brookes should touch him, but winked to others to do it, which two or three successively did, but the boy appeared not concerned. The justice then called on the father to take him, but had

* Daimonomagia, a small Treatise of Sicknesse and Diseases from Witchcraft and Supernatural causes, never before published, at least in this compressed order and general manner, being useful to others besides physicians, in that it confutes Atheistical, Sadducistical, and Sceptical Principles and Imaginations. London, J. Dover, 1665.

privately before desired one Mr. Geoffrey Strode to bring Jane Brookes to touch him, at such a time as he should call for his father, which was done, and the boy immediately sprung out after a very odd and violent fashion. He was afterwards touched by several persons, and moved not; but Jane Brookes being again caused to put her hand upon him, he started and sprung up twice, as before. All this while he remained in his fit, and some time after, and being then laid on a bed in the same room, the people present could not for a long time bow either of his arms or legs."

This boy, in his fit, appears also to have possessed the faculty of clairvoyance: he would describe, not only Brookes, but her sister, named Alice Coward. "He would tell," says the report, "the clothes and habits they were in at the time exactly, as the constable and others have found them on repairing to them, although Brookes' house was a good distance from Jones'; this they often tried, and always found the boy right in his description." It may be added, that to the deposition of these facts the two justices annexed the following attestation:—"The aforesaid passages were, some of them, seen by us, and some other remarkable ones, not here set down, were, upon the examination of several witnesses, taken upon oath before us. (Signed) Robert Hunt, John Carey."*

About the same period, Elizabeth Style, of Bayford, was indicted for bewitching a girl named Elizabeth Hill, thirteen years of age, and the deposition of three credible witnesses attests that during her fits "her strength was encreased beyond the proportion of nature, and

* Sadducismus Triumphatus, a Full and Plain Evidence concerning Witches and Apparitions. By Joseph Glanvil, Chaplain in Ordinary to King Charles II. London, 1726.

the force of divers men; furthermore, in one fit she foretold when she would have the next, which happened accordingly."*

A woman of the name of Florence Newton was committed to Youghal prison, 24th March, 1661, charged with bewitching Mary Longdon. In this case, during the trial, it is stated that when the accuser had closed her evidence the prisoner at the bar looked at her, and made certain motions of her hands towards her, upon which "the maid fell into most violent fits, so that all the people that could lay hands on her could scarcely hold her."† Another unfortunate woman, named Jane Wenham, charged with bewitching Anne Thorn, Anne Street, and others, was tried at the assizes of Hertford, March 1711-12, and received sentence of death. The case was tried before Sir Henry Chauncy, when the grand jury found her guilty of felony and witchcraft, and fully committed her to take her trial. Before the grand jury the depositions of sixteen witnesses were taken, one of whom deposed "that Jane Wenham confessed to him that she had practised witchcraft sixteen years."‡ On one occasion, when the girl she afflicted was in one of her paroxyms, we are informed that "a very ingenious gentleman and able physician happened to be present, his curio-

* Sadducismus Triumphatus, &c. p. 295.

† Witchcraft further Displayed. London, printed for E. Curl, at the Dial and Bible.

‡ A Full and Impartial Account of the Discovery of Sorcery and Witchcraft practised by Jane Wenham, of Walkerne, in Hertfordshire, upon the bodies of Anne Thorn, Anne Street, &c. Also her Tryal at the assizes, before Mr. Justice Powell. London, printed for E. Curl, at the sign of the Dial and Bible, against St. Dunstan's Church, Fleet Street, 1712.

sity bringing him a little out of his way to inquire into the truth of the story of this witch, which he had heard several ways told, as things of this nature generally are. When he saw her in a fit, which was one of the least she ever had, he tried whether he could bring her out of it without prayers. He took a great feather, which burning, he held under the maid's nose, and though the stink was so great that we were not able to bear it in the room, yet the maid received the strong steam into her nose without being the least affected by it, and without perceiving it, as far as we could perceive." The physician then felt her pulse, and assured them "it was no natural disease under which the maid laboured; that it must be counterfeit or preternatural;" "but," observes the author of this account, "that she should counterfeit even death itself one minute, and restore herself to health the very next, and that she should put herself to all this trouble for no manner of pleasure or profit, is so very inconceivable, and so wholly unaccountable, that I must needs say I shall never have faith enough to believe such a heap of absurdities."*

The insensibility of this girl in her fits was, it is stated, proved by M. Chauncy, who "ran a pin into her arm six or seven times, and finding she never winced for it, but held her arm as still as if nothing had been done to it, and seeing no blood come, he ran it in a great many times more, still no blood came, but she stood talking and never minded it; then again he ran it in several times more; at last he left it in her arm, that all the company might see it run up to the head."† The influence of direct manual appo-

* Discovery of Sorcery and Witchcraft, &c., p. 33.

† Ibid., p. 19.

sition in aggravating or calming the fits, is, in a variety of these cases, very distinctly attested. "There are also some things," continues the narrator, "in which the fits of Mary Longdon and Anne Thorn agree, particularly the great strength of the afflicted when in a fit,—so great, that three or four men could hardly hold 'em down; but there is one very remarkable difference which I doubt not my readers have already taken notice of,—viz. that this Mary Longdon was always worse of her fits whenever Florence Newton came into the room, whereas Anne Thorn constantly recovered from hers at the touch of the witch. And yet I think these different appearances may be accounted for different ways. It is not reasonable to suppose that either of these alterations in the afflicted came to pass by the consent or procurement of the witches themselves, who could not but perceive that they served as strong circumstances against them; but this was done by the overruling providence of Almighty God, to convict these miserable creatures, and either of these ways might do as well as the other, since it is equally surprising to see one in perfect health fall into such terrible fits at the sight of any one person as to see another recover out of such fits by the bare touch of the suspected witch, both of them tending only to the discovery of the criminal."*

It is also observed by Increase Mather, "As for that which concerns the bewitched persons being recovered out of their agonies by the touch of the suspected party, it is various and fallible; sometimes the afflicted person is made sick instead of being made whole by the touch of the accused; sometimes the power of imagination is such as that the touch of a

* Discovery of Sorcery and Witchcraft, &c., pp. 17, 18.

person innocent and not accused shall have the same effect." Bodin relates that "a witch who was tried at Nantes was commanded by the judges to touch a bewitched person, a thing often practised by the judges of Germany in the imperial chamber. The witch was extremely unwilling, but being compelled by the judges she cried out, 'I am undone,' and as soon as ever she touched the afflicted person the witch fell down dead." "I think," continues Mather, "that there is weight in Dr. Cotta's argument,—viz. that the power of healing the sick and possessed, was a special grace and favour of God for the confirmation of the truth of the Gospel; but that such a gift should be annexed to the touch of wicked witches, as an infallible sign of their guilt, is not easy to be believed. It is a thing well known, that if a person possessed by an evil spirit is (as oft it happens) never so outrageous whilst a good man is praying with and for the afflicted, let him lay his hand on them, and the evil spirit is quiet."*

In the year 1696 a commission was appointed in Scotland, by the Lords of his Majesty's Privy Council, to inquire into the case of Christian Shaw, daughter of John Shaw, of Bargarran. A quorum of these commissioners being met at Bargarran, and the accused persons confronted before Lord Blantyre, the rest of the commissioners, several other gentlemen of note, and ministers, the accused and, in particular, Catherine Campbell, were examined in the presence of the commissioners. "When they [the accused] severally touched the afflicted girl," says the report, "she was seized with grievous fits, and cast into intolerable agonies; others then present did also touch her, but no such effects

* Cases of Conscience, *loc. cit.*

followed: and it is remarkable that when Catherine Campbell touched the girle, she was immediately seized with more grievous fittes, and cast into more intolerable torments than upon the touch of other accused persons, whereat Campbell herself being daunted and confounded, though she had formerly declined to bless her, uttered these words:—'The Lord of heaven and earth bless thee, and save thee, both body and soul.'"*

During these trials we are informed that the prisoners were called in one by one, and placed about seven or eight feet from the justices, and the accusers then stood between the justices and them. "The prisoners were ordered to stand right before the justices, with an officer appointed to hold each hand lest they should herewith afflict them; and the prisoners' eyes must be constantly on the justices, for if they looked on the afflicted they would either fall into fitts or cry out they were much hurt by them.†" "On the trial of Bridget Bishops," it is further added, that "the indictment being drawn up according to form, it was testified at the examination of the prisoner, before the magistrates, that the bewitched were extremely tortured. If she did but cast her eye on them they were presently cast down,

* Sadducismus Debellatus, or, A True Narrative of the Sorceries and Witchcraft exercised by the Devil and his Instruments upon Mrs. Christian Shaw, in the county of Renfrew, in the West of Scotland, from August 1696 to April 1697; containing the Journal of her sufferings, as it was exhibited and proved by the voluntary confession of one of the witches, and other unexceptionable evidence, before the commissioners appointed by the Privy Council of Scotland to inquire into the same. Collected from the Records. London, printed for H. Newman and A Bell, at the Grasshopper in the Poultry, 1698.

† Another Brand Plucked out of the Burning, or, More Wonders of the Invisible World. London, 1700.

and this in such a manner that there could be no collusion in the business. But upon the touch of her hand upon them when they lay in their swoons, they would immediately revive, and not upon the touch of any one else. Moreover, upon the special actions of her body, as the shaking of her head, or the turning up of her eyes, they presently fell into the same postures; and many of the like accidents fell out while she was at the bar."*

In the year 1697, Richard Dugdale, a boy, nineteen years of age, excited considerable attention in Surrey as a demoniac; his fits were witnessed by numerous clergymen, physicians, and persons of respectability, from whose evidence it is perfectly clear that he was repeatedly in a state of magnetic extase. His fits commenced with violent convulsions, his sight or eyeballs turned upwards and backwards; he afterwards answered questions, predicted during one fit the period of accession and duration of another fit, spoke in foreign languages, of which at other times he was ignorant, and described events passing at a distance. Here again I shall quote *verbatim* the words of the narration: "At the end of one fit the demoniac told what hour of the night or day his next would begin, very precisely and punctually, as was constantly observed, though there was no equal or set distance of time between his fits; betwixt which there would be, sometimes a few hours, sometimes many; sometimes one day, sometimes many days." "He would have told," says one of the deponents on oath, "when his fits would begin, when they were two or three in one day, or three or four days asunder, wherein he never was, that the deponent knoweth of,

* Ibid. p. 121.

disappointed. On one occasion, while the minister was preaching to him, he exclaimed, "at ten o'clock my next fit comes on." "Though he was never learned in the English tongue, and his natural and acquired abilities were very ordinary, yet when the fit seized him he often spake Latin, Greek, and other languages very well." "He often told of things in his fits done at a distance whilst those things were a-doing,—as for instance, a woman being afraid to go to the barn though she was come within a bow's length of it, was immediately sent for by the demoniac, who said, "Unless that weak-faithed jade come my fit will last longer." Some said, let us send for Mr. G.; the demoniac answered, "He is now upon the hay-cart," which was found to be true. On another occasion he told what great distress there was in Ireland, and that England must pay the piper. Again, one going by him to a church meeting, was told by the demoniac in his fit, "Thou needest not go to the said meeting, for I can tell thee the sermon that will be preached there;" upon which he told him the text, and much of the sermon that was that day preached. Lastly, it is certified by two of the deponents that "the demoniac could not certainly judge what the nature of his distemper was, because when he was out of his fits he could not tell how it was with him when he was in his fits."*

The records of these trials for witchcraft are replete with curious details, which will doubtless excite the incredulity of many persons; but let us distinguish between the facts themselves, as established by positive evidence, and the superstitious ideas which

* The Surey Demoniack; or, an Account of Satan's strange and dreadful Actings in and about the body of Richard Dugdale, of Surey, near Whalley, in Lancashire. London, 1697.

may accompany the report of them. I wish, indeed, that they should be divested of all ideal colouring, and viewed merely as abstract realities, the existence of which, all persons, of whatever rank or education, are capable of determining. That some impositions may have been practised, and those too successfully, I grant; but it is well, albeit somewhat quaintly, argued, by an old English author, that "frequency of deceit and fallacy will warrant a greater care and caution in examining, and a greater scrupulosity and shyness of assent to things wherein fraud hath been practised or may in the least degree be suspected; but to conclude that because an old woman's fancy hath abused her, or some knavish fellow hath put tricks on the ignorant and timorous, that therefore whole assizes have been deceived in judgment upon matters of fact, and that numbers of persons have been forsworn in things wherein perjury could not advantage them,—I say, such inferences are as void of charity as of good manners. ....In things of fact, the people are as much to be believed as the most subtle philosophers and speculators, since then sense is the judge; but in matters of notion and theory they are not at all to be heeded, because reason is to be the judge of these, and this they know not how to use."*

In Lancashire, seven persons in one family were, in the year 1600, reputed to be possessed, and exhibited all the signs of what was considered demoniacal possession. Without entering into the details, which, after the particulars already given of other cases, might be tedious, I shall refer only to a summary of the symptoms they exhibited. "It is worthy to bee remarked," says George

* Glanville, Sadducismus Triumphatus, p. 20—37

More, "that though these possessed persons had every one something peculiar to herselfe which none of the rest did shew, and that so rare and straunge that all the people were obliged to confesse it was the worke of an evil spirit within them; so had they many things in common, and were handled, for the most part, in their fittes alike..... They had all and every one very straunge visions; they heard hideous and fearful voices of spirits sundrie times, and did make marveilous answers back againe .... they were in their fittes ordinarilie holden in that captivity and bondage that for an houre, two or three, and longer time, they should neither see, heare, nor taste nor feel nothing but the divells, they employing them wholly for themselves, vexing and tormenting them so extreameley as that for the present they could feel no other paine or torture that could bee offered, no, though you should plucke an ear from the heade or an arm from the bodie .... They had also a marveilous sore heaving as if their hearts would burst, so that with violent straining some of them vomitted bloude many times .... They were all of them very fierce, offering violence both to themselves and others, wherein they shewed verie greate and extraordinarie strength. They were out of their right mindes, without the use of their senses, especially voyd of feeling, as much sense in a stock as one of them, or as possible in a manner to quicken a dead man as to alter or chaunge them in their traunces in anything they either saide or did ....They in their fittes had divers parts and members of their bodies so stiffe and stretchd out, as were inflexible or very hard to be bended...........They shewed very great and extraordinarie knowledge, as may appeare by the straunge things saide and done by them, according to that which we have already set down in the particulars ....They ever after their fittes were as well as might be, and very little or no hurte at all, although they had been never so sore tormented immediately before."*

Increase Mather, also, after giving an account of the Tryals of the New-England Witches, gives the following summary of the phenomena they exhibited. He assures us "That they did in the assembly mutually cure each other even with a touch of the hand, when strangled and otherwise tortured, and would endeavour to get to their afflicted companions to relieve them:"—that they did also foretel when another fit was coming on, and would say, "Look to her, she will have a fit presently;" which fell out accordingly, as many can bear witness that heard and saw it:—that the motions in their fits are preternatural as to the manner, which is so strange as a well person could not screw their body into; and the violence also is preternatural, being much beyond the ordinary force of the same person when they are in their right mind: —that the eyes of some of them in their fits are exceeding fast closed, and if you ask a question they can give no answer, and I do believe they cannot hear at that time; yet do they plainly converse with the appearances as if they did discourse with real persons."†

* A Briefe and True Discourse, contayning the certayne possession and dispossession of seven persons in one familie, in Lancashire, by George More, Minister and Preacher of the Worde, and now (for bearing witnesse unto this, and for justifying the rest) a prisoner at the Clinks, where he hath continued almost for two yeares. 1600. To this title is annexed the following notice:—"This discourse, Reader, was dispatched in December last. The difficultie of printing hath hindred the publishing thus long."

† A Further Account of the Tryals of the New-England Witches.

Accordingly it is evident that the symptoms of supposed demoniacal possession correspond precisely with the phenomena which characterise the magnetic state. How these effects were produced, whether by exciting the imagination, or the fears, or otherwise affecting the nervous systems of the afflicted, is not the question which is now at issue; all we have to do with is, the simple fact that such phenomena really were developed, that the report of them is not false, that they were not feigned, but were veritable effects, depending on the operation of causes which were not then, and may not yet be, clearly understood. That they are referable to such fixed principles is certain, and may be inferred from the very circumstance of their constant uniformity,—that is to say, these symptoms of possession have been alike in all parts of the world, although it is manifest there could be no collusion or contrivance between the parties which exhibited them, whereby any such agreement could be simulated. In France, for example, during the administration of Cardinal Richelieu, the nuns in the Ursuline convent, in the city of Loudun, were supposed to be possessed. They were all violently convulsed; they exhibited a preternatural degree of strength; they all, too, during their fits, displayed extraordinary knowledge, and conversed in various languages which they had never learned.* Information of these facts

...... To which is added, Cases of Conscience. By Increase Mather, President of Howard College. London, printed for J. Dutton, at the Raven, Poultry, 1693.

* " Quant aux signes de possession, ceux qui se font au corps et consistent en agitations et contorsions, il y en a tant et de si estranges que plus de trente médecins ont donné attestations qu'ils surpassent la nature........Les autres signes qui se font dans

being transmitted to Cardinal Richelieu, who, under colour of being minister, was at that time, *de facto*, king of France, he ordered the affair to be investigated, and Urban Grandier, against whom the cardinal entertained a secret animosity,* was accused of being the author of these effects, and formerly arraigned for the crime of sorcery. Twelve ecclesiastics were appointed his judges: the process lasted eight

les puissances intérieures et qui touchent l'esprit, sont en très grand nombre, et se multiplient tous les jours. Cinq cents personnes, sans parler du reste, entre lesquelles je mets les juges commis par le roi pour faire le procès à Grandier, je mets d'autres officiers de justice venus de divers endroits, des gentilshommes, des dames, enfin des personnes de condition et non suspectes, tesmoigneront que les possedées ont declaré les choses occultes, ont obéi aux commandements qui leur estoient faits en secret, et souvent elles estant absentes quand on leur commandoit, qu'elles respondent au Latin le plus difficile qui s'apprenne dans les escholes, qu'elles ont respondu diverses fois aux interrogations faites en Grec en Espagnol, Italien, Turc, Topinamboux, et de ces signes il y a des attestations."—Véritable Relation des Justes Procédures observées au fait de la Possession des Ursulines de Loudun et au Procès de Grandier. Paris, 1634, pp. 20, 21.

* The cardinal had more than one reason to dislike Grandier, and wish his downfal, for when he was no more than bishop of Loudun there had been a sharp dispute between them; but that which irritated him most against Grandier was, an information—which those who accused him of the crime gave the cardinal—that Grandier was the author of a libel entitled, "La Cordonnière de Loudun," which was a very severe satire on his person, birth, and pedigree.—Bayle. Dict. Hist. Art. Grandier.

The signs of possession required by the Roman ritual were three:—divination, understanding languages which they never learned, and preternatural strength of body. Dr. Seguin, in reply to Menage, attests that he heard these nuns of Loudun, in their fits, answer M. Razilli in the Topinamboux language; he adds, "I believe him more than myself, and I allege it to you because you know him to be a credible man."—Ibid., note B.

months, and after a very minute and, apparently, solemn inquiry, he was condemned to be burned alive, and to have his ashes scattered in the air. The sentence was executed in April, 1634; and during the most terrible crisis of his sufferings, he is reported to have conducted himself with heroic fortitude. "Now, whatever opinion," observes an English critic in a recent number of the Foreign Review, "we may entertain of the cardinal Richelieu, or of the twelve judges appointed by his directions, or of the Jesuits who were the confessors of these nuns, and the personal enemies of Grandier, we conceive that any one who chooses to read the details of the transaction, will find evidence enough to convince him that these convulsions did not originate in any conspiracy, on the part of the nuns, at least, against Grandier's life, but were in reality first caused by their excited imaginations, and their terrors of devils, and propagated by the contagion of imitation, which is known to be so powerful in such cases. Two of the exorcists themselves* died shortly afterwards in the belief that they were possessed, and shewed all the symptoms resulting from such a conviction."† Here I may again repeat, that we are dealing with the facts of the case, which the reviewer admits, and of which neither imagination nor imitation can afford any adequate explanation.

After the revocation of the edict of Nantes, the protestants, it is well known, were, throughout France, subjected to the most unrelenting persecution, and were massacred in great numbers. The peasants among the Cevennes mountains wandered about in

* The Fathers Lactantius and Tranquillus.

† Foreign Rev., article, Animal Magnetism, Nov. 1829, p. 122.

the open air, enduring every description of privation and misery; they held secret conferences for religious purposes, and many became excited into a state of ecstacy, which was characterised by all the phenomena of magnetic somnambulism. They were seized with vehement tremblings and convulsions; hence they were called the Trembleurs des Cévennes; they preached and prophesied, and spoke in a variety of foreign languages. It is remarkable that this epidemic prevailed principally among men; but women also were affected by it, and even children of a tender age.* Jurieu, in his Pastoral Letters, states that they had all the appearance in their ecstacy of being in a profound sleep; the sensibility of the body was completely lost, so that they were not even sensible to the pain of being burned: those who had only an imperfect knowledge of the French language in their ordinary state, expressed themselves with facility and purity; many prayed and prophesied in a wonderful manner; and on recovering, none of them recollected anything which had during the paroxysm occurred.†

Marshal Villars, who was commissioned by the government to appease these troubles, observes, "I have seen things of this sort, which I would not have believed had they not occurred under my own

* Bertrand gives us the following note from the deposition of Guillaume Brugnier,—"J'ai vu à Aubessaque trois ou quatre enfants, entre l'âge de trois et de six ans, saisis de l'esprit. Comme j'étais chez un nommé Jacques Boussige, un de ses enfans âgé de trois ans, fut saisi de l'esprit, et tomba à terre. Il fut fort agité, et se donna des grands coups de main sur la poitrine, disant en même temps que c'était les péchés de sa mère qui le faisaient souffrir." —Bertrand du Magnétisme Animal en France. Paris, 1826, p. 358.

† Ibid.

eyes ; all the women and girls of a whole town appeared to be possessed. They trembled and prophesied publicly in the streets. One had the boldness to prophesy before me for an hour. One of these prophetesses, twenty-seven years of age, was taken, about eighteen months ago, before the bishop of Alais, who interrogated her before several ecclesiastics. The creature, after having heard what he said, addressed him with a modest air, exhorting him no longer to torment the true children of God. She then addressed him for an entire hour in an uncouth language, just as we have formerly seen the Duc de la Ferté, when he had drank a few glasses, talk English before the people of that country. .... This girl talked in a style like him, both Greek and Hebrew."*

Another set of religious enthusiasts, who have been designated the Convulsionnaires of St. Médard,† exhibited phenomena of the same description at the tomb of the abbé Paris. These, and many other facts which might be adduced of the same description,

* Vie du Maréschal Duc de Villars, tom. i. p. 325.

† The abbé Paris was a gentleman of very good family of the robe, and eldest son of a counsellor of parliament in Paris. He dedicated his life to the church; scarcely ever moved from before his crucifix; submitted to the severest penances, and acquired among the Jansenists so great a reputation for sanctity, that when he died these sectarians believed that the sick, by praying at his tomb, could be restored to health. Hume himself affirms, that the extraordinary cures wrought in France upon the tomb of the abbé Paris "were immediately proved on the spot, before judges of unquestioned integrity, and attested by witnesses of credit and distinction, in a learned age, and on the most eminent theatre in the world."—Hume's Works, vol. ii., Essay on Miracles, p. 133. London, 1777.

clearly prove that enthusiasm, or fanaticism, may lead to the development of effects which are truly somnambulic.*

To return, however, to the trials for sorcery and witchcraft which took place in England; it is, as Glanvil has affirmed, not reasonable to suppose that so many witnesses would have perjured themselves in attesting facts, the falsity of which, had they so forsworn themselves, could so easily have been detected; besides which, it should be remembered, that the deposition of every witness was corroborated by other

* The miracles of prince Hohenlöhe fall under the same category, as also those extraordinary exhibitions of the disciples of Irving, which lately attracted considerable public attention. In one of those amusing dialogues in Fraser's magazine, entitled "Oliver Yorke at Home," in which series of articles we find much profound mental philosophy gracefully and clearly propounded, M. Heraud states, that he witnessed one of these girls, during the time she was in a state of ecstacy, extemporising a poetical rhapsody, which, says he, "was good in composition, and sound in theology. It had a remarkable resemblance to the religious Odes of Klopstock; his emphatic repetitions; his naked enthusiasm; but none of his originality, the phrases being, without exception, borrowed from the poetic parts of scripture."—Fraser's Magazine, vol. iv., Oct. 1831, p. 368.

M. Heraud has furthermore favoured the translator with the following particulars:—"On my entering the room, the girl I have designated as a prophetess, seemed a young woman of considerable physical force, and of rather a masculine size and stature. She was in a rigid semi-recumbent attitude of ecstacy, with her mouth open, but lips motionless, and, as from her throat, a continued stream of sound issued. When the lyrical chaunt terminated, the patient remained in the same attitude, and during a prayer which Mr. Irving immediately offered, seemed to recover her self-possession. What I have lately seen of animal magnetism, pre-supposing there is no fraud or collusion, seems to me to explain the phenomena of Irving's prophetesses."—Vide also Fraser's Magazine, vol. iv. January 1832, p. 755.

collateral evidence. It is true that the law which existed against witchcraft does not prove the existence of witches;* but the evidence on these trials establishes, as clearly as human testimony can establish anything, that effects which in that age appeared supernatural, were really produced. "No man," says the venerable Baxter, "was more backward to condemn a witch, without full evidence, than Judge Hale;" unhappily, however, facts were too often insuperable, which, instead of causing the unfortunate victims to be led to the scaffold or the stake, ought to have been made the subject of serious scientific investigation. "Many of these poor creatures," says Reginald Scot, very truly, "had more neede to be relieved than chastised, and more mete were a preacher to admonish them, than a jailor to keep them; and a physician more necessary to helpe them, than an executioner, or tormentor, to hang and burn them."†

* This was an observation of the learned Selden; he remarks, "The law against witches does not prove there be any, but it punishes the malice of those people who use such means to take away men's lives; if one should profess that by turning his hat thrice, and crying buz, he could take away a man's life, though in truth he could do no such thing, yet this were a just law made by the state, that whosoever should turn his hat thrice, and cry buz, with an intention to take away a man's life, shall be put to death." —Seldeniana. The statutes against witchcraft and sorcery continued in France to be in force until Louis XIV. issued an edict restraining the tribunals of justice from receiving information of witchcraft, and in England until they were repealed by George II. stat. 9, c. 5.—Vide Blackstone's Commentaries. By Ryland, vol. iv. p. 60.

† The Discoverie of Witchcraft. By Reginald Scot. London, 1665. Preface.

CHAPTER X.

ROYAL TOUCH—CURES OF GREATRAK'S—PHYSIOGNOMICAL SUPERSTITIONS.

PYRRHUS, Vespasian, Adrian, Apollonius, and other persons of less eminence, cured, as we have already seen, diseases by the touch. In England this power was first assumed as a divine gift by Edward the Confessor; and his contemporary, Philip I. of France, soon proclaimed that he was endowed with the same faculty. Without entering into any *rationale* of the facts, it may be sufficient to observe, that many of their august successors assumed the same privilege, and certainly dispensed many royal blessings. The cures wrought by Charles V. are attested by his confessor and historian, Raoule of Presle, who especially bears testimony to his power in relieving chronic rheumatic affections. The historian Etienne de Conti, describes very particularly the preliminary ceremonies observed by Charles VI. before proceeding to the actual touch of the sick. Louis XIII. endeavoured to perform similar cures; hence arose the caustic remark made by the Duc D'Epernon, who being informed of the exorbitant power he had conferred on the Cardinal de Richelieu, exclaimed, "What! Louis, then, has only reserved to himself the power of curing the king's evil!" His successors, in accordance with this presumed royal privilege, continued the custom of touching scrofulous patients at the ceremony of their coronation. The unfortunate Louis XVI. when he was crowned, conformed with the ancient usage; and after touching the sick patients presented to him, pronounced the old verbal *formula*—"The king touches thee, may God cure thee."—"*Le Roi te touche, Dieu te guérisse.*"

In England, after the reign of Edward the Confessor, his successors, both kings and queens, were reputed to enjoy the same faculty; and the efficacy of the royal touch was for many centuries generally accredited. Bishop Tooker, chaplain to Queen Elizabeth, published, in 1597, a work entitled "Charisma Sive Donum Sanitatis," in which he attests the extraordinary ability of her majesty in curing, by the touch, persons afflicted with scrofula. William Clowes, Esq., one of the surgeons to her majesty, published, in 1602, another book entitled "Right, Fruitful, and Approved Treatise of Struma," in which he also compliments her majesty on effecting a cure which he judged to be "more divine than human," on a person who was suffering from scrofula. We meet about this time with the following anecdote:—"A papist, in prison, who had been sore troubled with the king's evil, having been cured by queen Elizabeth, after five years' experience from physicians in vain, being demanded—what news? I perceive, said he, that the excommunication against the queen can be of no avail, since God hath blessed her with such a gift."* Subsequent to this, John Bird, about the period of the restoration, published a treatise entitled "Ostenta Carolina," wherein he

* Wonders no Miracles. 4to. London, 1666.

maintains that Charles II. was the antitype of Edward the Confessor, and was able, by the royal touch, to cure, not only scrofula, but other diseases. About the same period, Dr. Thomas Allen published a work entitled the "Excellency, or Handy Work of the Royal Hand," wherein, in allusion to Valentine Greatrak's, who at that time was performing many remarkable cures by the application only of his hand, warns all persons against "the seventh sons, or those called strokers," whose success lessened the esteem of his majesty's performances.* The most eminent author, however, whose evidence may certainly be esteemed conclusive, is Wiseman, serjeant-surgeon to Charles II., who published several chirurgical treatises of very considerable merit. In his chapter on the cure of the king's evil, he observes, "I myself have been a frequent eye-witness of many hundreds of cures performed by his majesty's touch alone, without any assistance of chirurgery; and these, many of them, such as had tried out the endeavours of able chirurgeons before they came thither. It were endless to recite what I myself have seen, and what I have received, acknowledgments of my letter not only from several parts of this nation, but also from Ireland, Scotland, Jersey, and Guernsey."†

The ceremony of conferring the royal touch was, during the reign of Henry VII., performed in the fol-

* A Free and Impartial Inquiry into the Antiquity and Efficacy of Touching for the Cure of the King's Evil. By William Beckett, Surgeon, F.R.S. London, 1722. p. 26, and *seq.*

† Eight Chirurgical Treatises. By Richard Wiseman, Serjeant-chirurgeon to King Charles II. London, 1734. Vol. ii. p. 393.

lowing manner:—the king, priest, physician, the sick, and others being assembled, his majesty first knelt down, saying, "In nomine Patris et Filii et Spiritus Sancti, Amen—Benedicte." Then the chaplain, with a stole round his neck, kneeling before the king, repeated a short prayer, to which the king responded: "Confiteor Deo, Beatæ Mariæ Virgini, omnibus sanctis," &c. The chaplain then, after another prayer, read a portion of the gospel in Latin, during which the clerk of the closet brought forward the sick person, and both kneeling down before him, the king laid his hands repeatedly over the sore, and then crossed it with a gold coin, which was afterwards worn suspended round the neck of the patient until the cure was perfected. The chaplain having finished the gospel, the surgeon then led the sick person from the king. The pieces of gold used on these occasions were not originally coined expressly for the purpose, but were the current coin of the realm, called Angel Nobles, and composed of the purest gold. In the reign of Queen Elizabeth the coin bore the inscription, "A Domino factum est istud, et est mirabile in oculis nostris," but after her reign, it being suspected that many persons applied to be touched for the sake of the gold only, the piece was reduced to a coin of much smaller size and value, bearing the inscription, "Soli Deo Gloria." It was controversially alleged that the cure, instead of being effected by the touch, depended on the amulet virtues of the gold; but, says Wiseman, "We have many evidences to the contrary, for his majesty's royal father, in his great extremity or poverty, had no gold to bestow, but instead of it, gave silver, and sometimes nothing; yet in all those cases did cure, and those that were cured by his blood wore no gold."*

However extraordinary it may appear in this age of incredulity, it is certain that by the royal touch many cures were wrought. "That divers persons," says Bishop Bull, "desperately labouring under the king's evil, have been cured by the mere touch of the royal hand, assisted with the prayers of the priests of the church, is unquestionable; unless the faith of all our ancient writers, and the consentient report of some hundreds of most credible persons in our own age attesting the same, were to be questioned."† "I, myself," observes Dr. Douglas, bishop of Salisbury, "have met with a man who assured me that Queen Anne's touch, and the medal she hung about his neck, (and which I saw about it above twenty years afterwards,) had cured him when in a most desperate way. But, as a single unnamed witness will perhaps meet with little credit, what will you say when I name one whose attestation must be owned to be free from every suspicion? I have in my eye the late Mr. Dicken, serjeant-surgeon to Queen Anne, a gentleman well known for eminence in his profession. That he had opportunities of knowing the truth must be allowed, for all who were to be touched were first examined by him before they were brought to her majesty, and he made no secret in bearing witness to the certainty of some of the cures. This attestation of his must be consistent with the knowledge of many still alive, and I

* Wiseman, *Op. cit.*

† Some Important Points of Christianity Maintained and Defended in several Sermons and Discourses. By G. Bull, D.D., Bishop of St. David's; edited by Robert Nelson. 3 vols. Oxford, 1816. Vol. i., p. 136.

relate it to you on the authority of a physician of the strictest honour and probity, who assureth me that Mr. Dicken hath declared to him that he had in his possession a great number of letters writ to him by persons of character and distinction from most parts of England, attesting the recovery of their relations, acquaintances, and servants, who had come up to London to be touched, and thanking him for the trouble he had taken; that he himself could, from his own knowledge, affirm that several who were touched received benefit; and, in particular, he related one case which I shall here mention. A woman came to him, begging that he would present her to be touched by the queen. As, from her appearance, he had no great opinion of her character, he told her that the touch would be of little service to her, as he supposed she would sell her medal, which should continue about her neck to make the cure lasting. She promised to preserve it, was touched, and the medal given to her; and soon afterwards her sore healed up. Forgetting her promise, and now looking upon the piece of gold as useless, she disposed of it; but soon after her sores broke out once more. Upon this she applied to Mr. Dicken a second time, entreating him to present her again to the queen. He did so, and once more she was cured."* In allusion to the sick relapsing under such circumstances, Wiseman, vindicating the divine power of the royal touch, observes, "Now, whereas, some of them, upon the loss of their gold, have found damage, I should like to know whether any of them were relieved by any other

* The Criterion, or Miracles Examined with a View to Expose the Pretensions of Pagans and Papists. London, 1754. 8vo. pp. 199, 200.

gold than what the king gave them."* That this sanative blessing, however, was not conferred by the hand of sovereignty alone is manifest, for in Germany the same faculty was ascribed to the Counts of Hasprug, and the Saludadores in Spain professed to heal all external sores by the touch and the application of white linen.

In various parts of Europe individuals have occasionally appeared who professedly cured diseases by manual contact. One of the most celebrated of these was Valentine Greatrak's, an Irish gentleman, of high respectability in the county of Waterford.† He appears to have cured a variety of diseases, by simply passing his hands over the surface of the body; and it is interesting to know that among the number of his patients were the second Villiers, Duke of Buckingham; the son of the philosopher Cudworth, and the astronomer Flamstead. His fame having spread throughout Ireland, he was requested, by the Earl of Orrery, to visit England, for the purpose of treating the Right Honourable the Viscountess Conway, who was afflicted with a severe chronic headache, which had resisted the skill of the most eminent physicians in France and England. He accordingly visited England in January 1666, and resided at Ragley, the seat of Lord Conway, in Warwickshire. He afterwards removed to Worcester, where his success was so great, that he received an order from Lord Arlington, secretary of state, to proceed to London. On his arrival there,

* Wiseman, *Op. Cit.*, p. 27.

† He was born 14th February, 1628. The orthography of his name appears somewhat varied; by Glanville, it is spelled Greatrakes; by Henry Stubbe, Greatarick; he himself signs it Greatrak's.

he took a house in Lincoln's-Inn-Fields, and, during a residence of many months, performed many remarkable cures, an account of which, with certificates from the parties, he published in a letter addressed to the Honourable Robert Boyle. "The cures therein narrated," says Dr. Douglas, "seem to have all the circumstances necessary to establish any matters of fact. The names of those who were cured, the time when, the places where, and the witnesses before whom the cures were performed; all these marks of a genuine narrative are to be found in his book, which was besides published on the spot where the facts in question lay, exposed to every one's enquiry, so that had there been any fraud it must have been discovered; but indeed no such thing was or could be affirmed. The witnesses, many of them at least, are persons of such unexceptionable credit, good sense, and learning, that as we cannot suspect they were imposed upon, so we cannot suspect they would impose. Besides the certificates of many gentlemen of distinction,* we have also the attestations of grave Divines,† and eminent physicians,‡ which last are not very ready in admitting that cures may be effected without making use of the medicines which they themselves prescribe."§

* The Honourable Robert Boyle, Sir Nathaniel Holbatch, Sir John Godolphin, Sir Abraham Cullen, Sir Charles Does Son, Colonel Weldon, Alderman Knight, &c.

† Dr. Rust, Dean of Connor, afterwards Bishop of Dromore, Dr. Cudworth, Dr. Whichcote, Dr. Wilkins, Bishop of Chester, Mr. Patrick, afterwards bishop, &c.

‡ Sir William Smith, Dr. Denton, Dr. Fairclough, Dr. Jeremiah Astel, and others.

§ Select works of the Right Reverend John Douglas, M.D., late Lord Bishop of Salisbury. By his lordship's nephew, the Rev. W. Macdonald, M.A. Salisbury, 1820. Criterion, p. 483.

This letter of Greatrak's, addressed to the Honourable Robert Boyle, is dated London, 8th May, 1666, and is written in a modest and unpretending style. He states briefly his genealogy, and the education he received; he then accounts for his commencing the practice of healing by manual apposition, in the following manner:—"About four years since, I had an impulse or strange persuasion in my own mind (of which I am not able to give any rational account to another), which did very frequently suggest to me that there was bestowed on me the gift of curing the king's evil, which for the extraordinariness of it I thought fit to conceal for some time; but at length I communicated this to my wife, and told her that I did verily believe that God had given me the blessing of curing the king's evil, for whether I were in private or public, sleeping or waking, still I had the same impulse; but her reply was to me, that she conceived this was a strange imagination."* He then adds, "that a few days afterwards he tried his skill on a man who was grievously afflicted with this disease, and within a month he was perfectly cured."........... "Within some small time after this, God was pleased, by the same or the like impulse, to discover unto me that he had given me the gift of healing, which the morning following I told my brother and wife, but neither of them could be pre-

* A Brief Account of Mr. Valentine Greatrak's, and divers of the strange Cures by him lately performed. Written by himself in a Letter addressed to the Honourable Robert Boyle, Esq. Whereunto are annexed the Testimonials of several eminent and worthy persons, of the chief matters of fact therein related. London, printed for J. Starkey, at the Mitre, in Fleet Street. 1666.

vailed upon to believe it, though, for my own part, I had a full assurance thereof within me. This impulse I had the Sunday after Easter day, the 2nd of April, '65, early in the morning; and the Wednesday ensuing I went to Cornet Deans (about some occasions I had with him), to Lismore, where there came into his house to me a poor man, that with a violent pain in his loins, went almost double, who desired me, for God's sake, that I would lay my hands on him, and do him what good I could." Accordingly, Greatrak's complied with his request, and succeeded in curing him. He does not, however, pretend that he was, in all cases, successful; on the contrary, he distinctly states that he was unable to relieve the Viscountess Conway of her painful malady, and resolved on returning home. "But," says he, "at the entreaty of my Lord Conway, I remained with his honour at Ragley, three weeks or a month, where many hundreds daily came to me, and many were cured of their diseases and distempers, and many were not."* Again he observes, "Many may demand of me why some are cured and not all? to which I make this answer, that God may please to make use of such means, by me, as shall operate according to the disposition of the patient, and therefore cannot be expected to be the like effectual in all. They also demand further, why some are cured at once coming and not all; and why the pains should fly immediately out of some and take such ambages in others; and why it should go out of some at their fingers, some at their toes, some at their noses, others their ears or mouths? To which I say, if all these things could

* A Brief Account, &c., p. 39.

have a plain and rational account given of them, then would there be no reason to account them strange. Let them tell me what substance that is which removes and goes out with so great expedition, and it will be more easie to resolve their questions; but till then let them be silent and admire the works of God, whose ways are past finding out, and whose majesty is not confined to time, manner, or measure."*

During the time Greatrak's was on this visit at Lord Conway's, Mr. Henry Stubbes, physician, at Stratford-on-Avon, was invited to Ragley, where he met Greatrak's, and had the opportunity of becoming personally acquainted with him. He describes him to have been a man "of a graceful personage and presence," and, "if my phantasy," he adds, "betrayed not my judgment, I observed in his eyes and meene a vivacitie and spritelinesse that is nothing common .... I inquired of him how he came by his miraculous gift of healing, and he replied it was the grace of God in him, which he understood first by a strong and powerful impulse in him to essay it, and this notion was so prevalent upon him that it incessantly runne in his minde, nor could any businesse, how serious or religious soever, divert him from those cogitations."† His method of operating appears to have been very simple. "I considered," says Dr. Henry Stubbes, "that there was no manner of fraud

* A Brief Account, &c., p. 32.

† The Miraculous Conformist, or an Account of severall Marvailous Cures performed by Stroaking of the Hands of Mr. Valentine Greatrick; with a Physicall Discourse thereupon, in a Letter to the Honourable Robert Boyle, Esq. By Henry Stubbe, physician, at Stratford-upon-Avon. Oxford, 1666.

in the performances, that his hands had no medicaments upon them, ... nor was the stroking so violent that much could be attributed to friction. I observed that he used no manner of charms or unlawful words; sometimes he ejaculated a short prayer before he cured any, and always after; he bade them give God the praise." It is asked by Mr. Colquhoun, after citing this passage—"Now where is the imposture in all this?" Certainly there was none, and he must be somewhat arrogant of his own critical sagacity who would repudiate such testimonies as the following:—

Dr. Rust, the Lord Bishop of Dromore, observes, "I was three weeks together with him at my Lord Conway's, and saw him, I think, lay his hands upon a thousand persons; and really there is something in it more than ordinary, but I am convinced it is not miraculous. I have seen pains strangely fly before his hands, till he hath chased them out of the body—dimness cleared, and deafness cured by his touch; twenty persons, at several times, in fits of the falling sickness were, in two or three minutes, brought to themselves, so as to tell where their pain was, and then he hath pursued it till he hath drawn it out at some extreme part.............He pretends not to give testimony to any doctrine; the manner of his operation speaks it to be natural; the cure seldom succeeds without *reiterated* touches; his patients often relapse; he fails frequently; he can do nothing where there is any decay of nature, and many distempers are not at all obedient to his touch. So that, I confess, I refer all his virtue to his particular temper and complexion; and I take his spirits to be a kind of elixir and universal ferment; and that he cures (as

D. M. expressed it) by a kind of sanative contagion."* The testimony of Dr. Whichcote also, cannot, I apprehend, be in any way impugned. He states in his certificate, "I have been present when sundry persons thus helped and relieved have, with an abundance of joy and chearfulness, given God thanks for the great benefit they had received by means of Mr. Greatrak's. I have been in company with persons of good understanding and credit, who have, of their own accord, declared and testified that they themselves have been eye-witnesses of the like good effects. Tis true he does not cure such in whom nature is spent, or where the parts or principles of nature altogether are defective, to whom, at first coming to him, he doth commonly declare, that he cannot help them : some others, at the first or second application to him, he doth not help, who afterwards are restored at their oftener coming to him...............For my own part he hath been to me an happy instrument of God to relieve and ease me of a very dangerous and painful malady, which for many years had greatly disabled and sorely afflicted me."† It does not appear that Greatrak's produced somnambulism, but then it should be remembered he did not comply with the elementary condition necessary to produce that effect; he did not operate with the intention of inducing it; however, many of his patients were convulsed, and some manifested that degree of physical insensibility which is so peculiarly characteristic of the magnetic state. In Dr. Wilkins' certi-

* Vide Letter to Glanville. Sadducismus Triumphatis, pp. 52, 53.

† Greatrak's Letter to the Honourable Robert Boyle. Certificates annexed.

ficate in reference to the case of a woman who had applied to Greatrak's to be cured of severe headache, it is stated, that, "after the operation, the upper joints of her fingers were dead, and that if she were wounded in them she would neither feel it nor would any blood come, for the tryal of which having first several times pinched the pulp of her fingers betwixt my nails, with great violence, I did several times thrust in a pin neer to the bone, but the woman was not sensible of either, neither did any blood come; but upon two or three slight strokes of his hand she professed to be eased of her pain, and her fingers were restored to their wonted temper, and then being but gently pricked with a pin, she was sensible of it, and the blood followed upon it. This I can truly and do willingly attest." This certificate is then signed by Dr. Wilkins, and countersigned by Mr. Sandys, who also witnessed the operation. Upon the whole, it will be found that the cures which Greatrak's performed by manual apposition are so fully and unequivocally attested by impartial and competent observers, that I do not think the most perverse and ingenious sceptic can, by any sophistry whatever, invalidate the evidence.

Many years after Greatrak's had acquired so much celebrity, John Joseph Gassner, the celebrated German exorcist, attracted considerable attention by the cures which he performed. He was born in 1727, at Bratz, near Pludenz in Swabia, and became catholic minister at Clösterle, in the bishopric of Chur. He began by healing his parishioners in an obscure town on the borders of Switzerland and Tyrol, and his reputation so much increased, that during the last two years of his residence there he had between four and five hundred patients under his treatment. He made a tour through several of the Swiss Cantons, and settled at Ratisbon, where he was visited by Mesmer, the celebrated Lavater, and many men of scientific eminence. His cures were performed with much pomp, and it was observed that he always rubbed his hands upon his girdle and handkerchief previous to his touching the patient. He performed the ceremony in the name of our Saviour, and by the faith of the diseased in his holy name; if the cure did not take place they wanted faith. He gave the sick, when he dismissed them, balm and oil, together with certain waters and powder, which they were to use as spiritual medicaments. A public officer took minutes of the cases he treated, which were attested by numerous persons of high rank and respectability.

In the darker ages, the art of effecting cures by such processes as the above was mixed up with the study of natural magic. Hence Hippocrates,* Dioscorides,† Theophrastus,‡ acknowledge the curative effects of magic, which was performed by looks, gestures, signs, and various ceremonious rites. Fascination and Enchantment are terms frequently used by old authors to express the magnetical power exerted by one person over another; and, like some modern theorists, Avicenna, Pomponatius, and many others who recognised the facts, ascribed them to the operation merely of the imagination. As regards the influence of physiognomical expression, the fascination or magical power of the expression of the eye has been always

* Hippocrat. de Sacro Morbo de Magis.

† Dioscorides, lib. ii. cap. 10.

‡ Theophrastus de Histor. Plant, lib. ix. c. 4.

a matter of popular observation.* Athenæus and Zustatius state, that one Cratisthenes was so excellent a magician that he could not only charm with the eyes, but change the very phantasies of men. The whole secret of fascination is described by Ficinus as depending on the witchcraft of the eyes. "Mortal men," says he, "are especially bewitched when, by often gazing one on another, they direct sight to sight —join eye to eye .... therefore he that hath a

* The fascination of the human eye, or the influence of one commanding spirit over another, is very finely described by Lord Byron, in Cain, where he makes Adah, on the approach of Lucifer, observe:—

"I cannot answer this immortal thing
Which stands before me; I cannot abhor him;
I look upon him with a pleasing fear,
And yet I fly not from him: *in his eye*
There is a fastening attraction which
Fixes my fluttering eyes on his; my heart
Beats quick; he awes me, and yet draws me near,
Nearer—and nearer. Cain,—Cain, save me." . . . . .

Cain, Act i. Scene 1.

So also in the rhyme of the Ancient Mariner, by Coleridge, the wedding guest is spell-bound by the eye of the old sea-faring man, and constrained to hear his tale:—

"He holds him with his skinny hand;
There was a ship, quoth he,—
Hold off, unhand me, greybeard loon,
Eftsoons his hand dropped he.
He holds him with his glittering eye—
The wedding-guest stood still,
And listens like a three years' child;
The mariner hath his will.
The wedding-guest sate on a stone,
He cannot choose but hear:
And thus spake on that ancient man,
The bright-eyed mariner."

clear eye, though he be otherwise deformed, will make one mad, and tie him fast to him by the eye."* So also Philostratus speaks of an Ephesian, who "by looking steadily on a person could kill him;" and Pliny mentions a whole family, "who," he says, "possessed so cross and felonious an aspect that they killed those on whom they gazed."† Furthermore, Plutarch affirms that some persons have such a power in their eyes that their friends are thereby fascinated,‡ and a variety of other passages from ancient authors might be cited to the same effect. The superstition of the "evil eye" is familiar to every one; among the Arabians it is yet entertained, and accordingly we find that Captain Lyon, in his travels in Northern Africa, informs us that he "observed, in a deep romantic valley, about six miles from Beni-Abbās, many of the trees ornamented with the skulls of horses, camels, or sheep, to protect them from being blasted by the evil eye."§

This, and other physiognomical superstitions, evidently derived their origin from the popular observation of the remarkable influence which some individuals almost unaccountably acquire over others; indeed the very sympathies which associate mankind together are inexplicable excepting on magnetical principles. The natural affinities which exist between chemical substances are, it is well known, determined by laws which are fixed and permanent; and so also the manifestations

* Apud Burtin. Anatomie of Melancholie.

† Plinius Hist. Nat., lib. vii.

‡ Plutarchus 5 Sympas, cap 7.

§ A Narrative of Travels in Northern Africa, in the years 1818, 1819, and 1820. By Captain G. F. Lyon. 4to. London 1821, p. 31.

of the human mind, and the affections of the human heart, are doubtless determined by laws equally inherent in the constitution of our nature; nay, to suppose otherwise, would be to destroy, imaginatively, the whole harmony of the moral world, for it would in reality be to presume the existence of a vast series of effects unproduced by any cause whatever, and existing independent of any laws to determine their mutual relations. Matter affects matter upon physical principles that are demonstrable, and mind, we may rest assured, affects mind on psychical principles, which ought to be more fully investigated.

CHAPTER XI.

EARLY WRITERS ON MAGNETISM—ANIMAL MAGNETISM IN ENGLAND.

In this great commercial country it is notorious that the attention of the government is so exclusively occupied in conducting its own domestic and foreign affairs, that science and literature do not receive the same immediate encouragement as in France. Here no royal commissions are delegated to inquire into any reputed great discovery; the author of it is left to fight single-handed against the opposition by which he must inevitably be assailed, and when he has succeeded in establishing the truth of his discovery, he is then not unfrequently honoured by the patronage and favours of the reigning monarch. So was it with Harvey, Sir Isaac Newton, Jenner; and so has it also been with many men of high literary distinction. This being the character of the government, under whatever administration it may be conducted, and there being no Institution like the Académie Royale de Médecine in Paris, or the Académie des Sciences at Berlin, to whose verdict I can appeal, I at once meet with this difficulty, that there is here no tribunal before which I can exhibit the facts I am prepared to demonstrate, whose decision might carry with it authority; for I am aware that the evidence of individuals, however in reality unimpeachable, may unjustly be impugned as insufficient; and yet it appears to me that the opinions in favour of animal magnetism, by such men as Cuvier, La Place, Ampère, Hufeland, Treviranus, Humboldt, Sprengel, Keiser, Brandis, Autentreith, Cloquet, Rostan, Georget, might at least disarm scepticism of its arrogance, and enforce on the conviction, especially of the medical profession, the propriety of giving the subject an impartial and candid investigation. What was the opinion of Dugald Stewart, who will not be accused of being a German visionary? It was in favour of animal magnetism. What was the observation of the acutely-minded scholar and philosopher, Coleridge? "I have seen that which I would not have believed on your report; and therefore, in all reason, I can neither expect nor wish you should believe on mine."* And what has been the conclusion of almost every scientific man who, either in France or Germany, has paid attention to the phenomena? It has been so decidedly in favour of the science that the existence of the facts is no longer, on the continent, a matter of dispute.†

* Coleridge. Table Talk, vol. i. p. 107.

‡ A brochure, entitled "Hints on Animal Magnetism, addressed to the Medical Profession in Great Britain," has just appeared, by the author of the "Isis Revelata." It is another earnest and energetic appeal to the medical profession in England to examine the phenomena. He observes that in this country "the facts of animal magnetism, and their scientific importance, may be considered as merely an affair of time. It required half a century upon the continent; but if the subject were once seriously taken up by competent inquirers amongst ourselves, a much shorter period ought to suffice; because we have it in our power to avail ourselves of all the previous researches of the continental physiologists. In France and Germany, where the investigation hitherto has been principally carried on, the facts in question, I believe, are now *generally recognised by philosophers, and reasoned upon, as*

We have already seen that many of the more ostensible phenomena of somnambulism obtruded themselves, in a manner, on the attention of the ancients; and we shall find that traces of the doctrines of Mesmer —"foregone conclusions"—founded on imperfect observation—are scattered through the works of many early authors. It is easy to denounce learned men, who lived in an age unillumined by the present light of science, as visionaries and impostors;—it is easy to raise an outcry of quackery against the most intelligent and zealous minded champions of truth; but the stream of human knowledge, although for a moment darkened by such prejudices, is not interrupted in its progression, or diverted out of its proper channel. The very vestiges of hostility and persecution are the landmarks which should warn men of science of a future age against falling into the same error; for what can be more irrational than to reject, without investigation, facts which are *ex facie* well authenticated? The early writers on animal magnetism imperfectly anticipated many of the doctrines maintained by the magnetisers of the present day. It occurred to Paracelsus, that all animated beings were endowed with an occult power analogous to mineral magnetism, which was derived from the stars, and which he called *magnale*. He insisted on faith as the elementary foundation of all the occult sciences. "The imagination," says he, "is confirmed and developed by faith; it must receive its full development before it can be brought to act, for the slightest doubt crushes the whole offspring;—

authentic elements, in almost all physiological and psychological works. What is true, in this respect, upon the continent, cannot be false in Great Britain. Science is of no particular country, and acknowledges no geographical boundaries."

faith strengthens imagination—faith determines the will; he who believes in nature receives from nature everything in proportion to his faith."* One of the commentators on Paracelsus, Leon Suavus, who appears to have studied magnetism, observes "All the phenomena of the will are not incredible to those philosophers who perfectly understand the virtues and character of the human mind, which is restrained only by the interposition of the body being in every other respect equal to that of angels. We pass over, without observation, fascination and other modes by which the mind exhibits outwardly very surprising phenomena. Nothing is more effectually productive of these marvellous effects than the imagination of one who places the most unbounded confidence in the object of his pursuit, whether it be real or ideal."† After Paracelsus, a certain Rumelius Pharamond, the Chevalier Digby, Crollius, Bartholin, and Anman, published treatises on the magnetic doctrine, and their opinions having found their way into France, were warmly espoused by Loysel, Dolé, Gaffarel, and some other men of learning.

In the beginning of the fourteenth century, Arnauld de Villeneuve, being well versed in the literature of the Arabs, drew from this source the doctrine of magnetism, and practised it in the treatment of diseases. The signs he adopted were soon reputed magical. He practised at Montpellier, where he was severely attacked by his colleagues, and anathematised by the Sorbonne. Harassed by misfortune, with his health shattered, and his spirits broken, he gave way to melancholy; and, like J. J. Rousseau, fancied himself

* Paracelsus Opera Omnia, Lat. Germ. fol. Strasb. 1603. fol. 110.

† Paracel. Comment. lib. i. p 236, fol. 1605.

an object of hatred to all mankind. After him appeared Pomponatius, so renowned for his wit, and particularly for the persecution he endured on account of his metaphysical opinions. His work on incantations, "De Incantationum Occultâ Potestate," had great publicity. He declares in it, that he does not believe in magic; but he affirms that the virtue of curing diseases possessed by certain men, is inherent in them, and that they can effect cures by touch, without any miraculous or magical intervention. He considers it possible that "health may be produced from without by a soul imagining as well as desiring it." He also states, "that his opinion is not the same as that of the Arabs. According to Avicenna, the soul never acts but from her own independent consciousness, and by the sole force of her will; but, according to him, the soul never acts but by attraction, and through the vapours which she propels towards the patients." "There is in the mind of man," says he, "a certain power of changing, attracting, checking, and binding men and things according to his desire; for everything yields to him when he is carried to an extreme of passion or virtue; but only inasmuch as he excels those whom he wishes to bind."

Cornelius Agrippa, born at Cologne in 1486, carried his ideas of the extension of faith and imagination very far. "The passions of the mind which are led," he observes, "by fancy, when they are very strong, can not only change the body of their possessor, but can also influence the bodies of other persons, and even communicate or cure diseases both mental and physical." He also remarks, and the idea involves an excellent moral principle, that "some men influence you by their sole affection, the peculiar habit which surrounds them in consequence of which we should avoid all intercourse with the wicked and the wretched; for their soul radiates a noxious influence, contaminating those who come near them with a sort of contagious wretchedness; for the same reason, the case being reversed, he recommends the society of good persons."

At a later period, Nicolas of Lucca wrote a treatise on the magnetism of the blood; Laurence Strause, another on magnetic sympathy; and Pierre Borel, the king's physician, and a member of the "Académie des Sciences," in a dissertation published by him, maintains, on sympathetic cures, the same doctrine, and admits, not only the influence of the general fluid, but likewise that of the will on the animal economy.

In the year 1608, Goclenius, professor of medicine at Marburg, published a rather long treatise on magnetic cures, in which he endeavours to prove that such cures are effected by a natural process which he endeavours to explain. In 1621, Vanhelmont, a disciple of Paracelsus, published his treatise on the magnetic cure of wounds, a work in which he defends his teacher against the attacks of several Jesuits. The inquisitors, astonished at the profound knowledge of medicine possessed by Vanhelmont, regarded him as a magician, and ordered him to be arrested. He escaped, however, from the dungeons of the inquisition, and went, like his cotemporary Descartes, to seek peace and liberty in Holland. Here he published, at the Elzevir establishment, his work entitled, "Effects of Magnetism on the Human Body," in which he developed many new and striking views. "Magnetism," says he, "exerts its influence every where; there is nothing new in it, excepting the name; it seems a paradox to those only who deride everything, attributing to the evil spirit whatever they cannot explain. Supposing a witch practise witchcraft, it is not through the agency of the devil, for he cannot communicate a power which he does not possess; it is by means of a faculty peculiar to, and inherent in, human nature, and which, like every other faculty with which we are endowed, may be applied either to a good or to an evil purpose."

According to Vanhelmont, the soul is endowed with a modelling power, which, after having produced an idea, can mould it into substance and shape, and send it far off, propelled and guided by the will. This power, infinite in the Creator, is, of course, limited in his creatures, and may consequently be more or less obstructed by obstacles. Ideas thus clothed in substance, have a physical influence on living beings through the medium of the vital principle; they act with more or less power, in proportion to the energy of the will that sends them forth; and their action may be checked by the resistance of the receiving patient. Some learned men, in Vanhelmont's time, thought that the sympathetic power emanated from the stars, but Vanhelmont maintained the contrary. "I see," said he, "the source of this power in an agent much nearer to ourselves. Ideas are the directing power, and these ideas are produced by charity or good-will. Thus, respecting the sympathetic influence, I consider these stars of our intelligence (attention and charity) far above the stars of the heavens. Such ideas arising through a desire of doing good are heard afar off, and reach the object pointed out by the will, whatever may be the intervening distance."

It is impossible to have a more correct idea of the

nature of the magnetic agency; it is much to be regretted that such truths should ever have been mingled and confounded with the superstitions of the age in which they were promulgated. Vanhelmont was supported in his subsequent attempts to support this doctrine by Robert Flud, a Scotchman, who was the author of several works on philosophy.

According to Robert Flud, man has two poles, like the earth; and magnetism can only take place when his body is in the proper position. After having examined, on this point, the opinions of various authors, chiefly those of Plato, Pythagoras, Aristotle, and Empedocles, he concludes that, besides the action of the poles in man, which he admits as demonstrated, there exist two principles acting continually upon him, which he thus describes:—" If, when two persons come together, the magnetic rays or emanation they send forth, happen to be repelled, reflected, or repercussed, from the circumference towards the centre, *antipathy* exists, and magnetism is negative. If, on the contrary, there is abstraction on either side, and a free emission from the centre to the circumference, *sympathy* takes place, and magnetism is then positive." He admits several species of magnetism; and the evidence he adduces in support of his assertions, is extremely curious.

Robert Boyle, the founder of the Royal Society of London, a profound mathematician and learned physician, perceived the action and the reaction which individuals could exert upon one another; and he admitted a general fluid productive of this result. In his highly esteemed treatise on the corporeal affluences, proving their extreme subtility, he also established their power and influence.

The same researches had already been made by the Spaniard, Balthazzar-Gracian. Attraction, which he seems to have known, is called by him, with much propriety, the natural relationship of the minds and hearts of men. He perceived this permanent flux and reflux of the vital principle and corporeal humours in man, without which he thought the motion necessary to life would be interrupted; and he explained the effects of sympathy and antipathy in such a manner as to make them lose much of their mystery. "The atmosphere peculiar to each individual," says he, "retains as much of the attraction and repulsion of the general fluid, as the nature of his constitution allows. In the various intersections of these individual atmospheres, some emanations are more attractive between two beings, others more repulsive, than others." This doctrine differs but little from that of the ancient Rabbin Abraham Benhannas. "The magnet," said he, "attracts iron; iron is found everywhere; everything therefore is under the influence of magnetism. It is only a modification of the general principle which establishes harmony, or foments divisions among men; it is the same agent which gives rise to sympathy, antipathy, and the passions." Father Kircher, whose work contains an account of that of Robert Flud, says, towards the conclusion, that his book must have come from the devil; but he goes much further than his predecessors in adducing known instances of sympathy and antipathy, and every species of affinity observable in nature; all of which appeared to him so many species of magnetism. He distinguishes several kinds,—namely, the magnetism of metallic bodies, that of the sun, that of the sea, that of animals, which he denominates "*zoo-magnétismos*." Kircher was so possessed with this sole idea, that all nature appeared to him like one immense magnet.

Another learned author, Jean-Baptiste Porta, whose house was the resort of all the *literati*, and who founded at Naples the Academy of Secrets (*de secreti*), so called because, in order to be admitted a member, it was indispensable to be the inventor of some secret or discovery, adopted part of Kircher's opinions, and published a book on the corpuscular philosophy, containing several magnetic prescriptions, and a treatise on natural magic. The cures operated by Porta appeared so extraordinary to the court of Rome, that, on the suspicion of his being a magician, he was interdicted, and his cures prohibited.

Others, especially, Coclius and Porta, confounded the science of magnetism with the study of astrology; the former predicted the nature of his death, and apprized his murderer, at Caponi, that ere the sun went down he would commit a crime, which proved true, for he was murdered by that very man; the latter foretold Spinola, the Genoese, that he would meet with great success if he would embrace the military profession, which happened according to the prediction.

Wirdig, professor of medicine at Rostoch, persuaded himself that there was in nature, and its bodies, more life, more motion, more magnetism, more intelligence, than had been heretofore admitted. He enlarged the system of Kepler; his own system appeared under the title of "New Medicine of the Spirits;" he presented it to the Royal Society of London, where it was printed in 1673. According to Wirdig, the magnetic influence takes place, not only between the celestial and the terrestrial bodies, but this influence is reciprocal. "The whole world," say she, "is submitted to the power of magnetism. Life is preserved by magnetism ; everything perishes by magnetism."

Stahl, to whom chemical science is indebted for much of its lustre and glory, discovered among the ancients the doctrine of magnetism, and made novel and felicitous applications of its principles. His "True Medical Theory," printed in 1708, announced the grasping genius with which he had studied all the branches of his art. It contains some acute observations, founded on the existence of a vital principle circulating in all beings, by which their actions are modified, and governed, and which he supposed possessed, like all currents, a sort of flux and reflux, the absence of which in man produces obstructions, paralysis, epilepsy, convulsive movements, and death.

Maxwel entertained the same ideas ; he reduced all his observations into a regular system, and devoted himself to the improvement of magnetic medicine, which he boasts of having first drawn out of its chaos. His work was published at Frankfort, and reduced into aphorisms by Ferdinand Santanelli. Among a variety of things it contains, unintelligible to those who are not familiarised with the study of occult powers, we meet with very ingenious ideas, explaining the influence exerted by one individual over another, and sometimes over his own organisation. "All bodies," says he, "emit corporeal rays serving as vehicles through which the soul transmits her influence, by communicating to them her energy and power of acting ; and these rays are not only corporeal, but they are even composed of various kinds of matter. When the spirit intimately connected with the properties of one body, communicates with another body, the mutual flux and reflux of the spirits from one body to the other, establishes a kind of sympathetic connexion, which is not so easily dissolved as when it is the work of imagination. The universal remedy is no other than the vital spirit strengthened in a suitable subject. According to the primitive design of nature, every subject receives the exact share of vital spirit necessary for its preservation after its kind ; it is, however, possible that nature, through the exertions of a philosopher, may be made to produce things superior to their own principle. If you wish to work prodigies, abstract from the materiality of beings, increase the sum of spirituality in bodies, rouse the spirit from its slumbers. Unless you do some one or other of these things,—unless you can bind the idea, you can never perform anything good or great."

Thus it is evident that these old authors imperfectly anticipated many of the views of Mesmer, but facts were as yet wanting to enable them to establish anything like a permanent system ; and these were not supplied until the discoveries of Mesmer and Puységur put men of science in possession of the method of inducing the magnetic state. It was soon found that there was no mystery, no secret art, no difficulty, in conducting the operation ; and facts therefore have, during the last fifty years been accumulated in every part of Europe ; so that the induction I apprehend is now sufficiently large, and complete, to warrant our coming to more satisfactory conclusions.

The discovery of animal magnetism was, in England, received with considerable distrust and apathy ; and as soon, therefore, as the unfavourable report of

Franklin, Lavoisier, Bori, &c., appeared in France, it was translated* into English, and appealed to as conclusive evidence against the doctrine. It was not then perceived,—nor is it now by many persons sufficiently understood,—that this very report attests the facts of animal magnetism. The commissioners dispute only the fluid theory of Mesmer; instead of being adduced, therefore, as an evidence against the system, it is in reality a conclusive attestation in its favour. In the year 1788, Dr. Maineduc, a pupil of Deslon's, arrived in England, and delivered a course of lectures on animal magnetism at Bristol,† and afterwards in London.‡ He also treated magnetically, and with considerable success, a great number of cases, an account of which, with certificates from the

* Report of Dr. Benjamin Franklin and other Commissioners, charged by the King of France with the Examination of Animal Magnetism as now practised in Paris. Translated from the French, with an Historical Introduction. London: Johnson, St. Paul's Church-yard. 1785.

† "On looking over the list of students," says Dr. George Winter, "that had been, or then were, [at Bristol] under the Doctor's tuition, it appeared that there was 1 duke, 1 duchess, 1 marchioness, 2 countesses, 1 earl, 1 lord, 3 ladies, 1 bishop, 5 right honourable gentlemen and ladies, 2 baronets, 7 members of parliament, 1 clergyman, 2 physicians, 7 surgeons, exclusive of 92 gentlemen and ladies of respectability; in the whole, 127."—History of Animal Magnetism. By George Winter, M.D. Bristol, 1801. 8vo. pp. 14, 15.

‡ Hannah More, in a letter dated Cowslip-green, September 1788, addressed to Horace Walpole, observes, "I give you leave to be as severe as you please on the demoniacal mummery which has been acting in this country. .... Mesmer has got a hundred thousand pounds by animal magnetism in Paris. Maineduc is getting as much in London."—Memoirs of the Life and Correspondence of Mrs. Hannah More. By William Roberts. London, 1834, vol. ii. p. 120.

patients themselves, he afterwards published in a pamphlet entitled "Veritas," which bears the appropriate motto, "*Causa latet, vis est notissima.*"* His lectures excited very considerable sensation in scientific and literary circles; and, as might be expected, a number of magnetic practitioners, in imitation of him, soon entered the field of competition. We are informed by Dr. George Winter, that a person named Holloway, by giving lectures on animal magnetism at five guineas for each pupil, realised a considerable fortune; and the house of Mr. Loughborough, another magnetic professor, at Hammersmith, about the year 1790, was daily for many months crowded with patients. "In the year 1790," says Dr. George Winter, "I deem animal magnetism to have been at its height; it was credibly reported that 3000 persons have attended at one time to get admission to Mr. Loughborough's, at Hammersmith, and that some persons sold their tickets for from one to three guineas each."† But, notwithstanding all this, while animal magnetism was making rapid progress in Germany and France, it does not appear to have made any advancement in England; on the contrary, the fanatical interpretation which a Mrs. Pratt put on the cures of Loughborough, could not fail to have disgusted many who might otherwise have been interested in the facts themselves, which were very clearly and unequivocally established.‡

* Veritas; or, a Treatise containing Observations on, and a Supplement to, the two Reports of the Commissioners appointed by the King of France to examine into Animal Magnetism, with Certificates, &c., of cures performed. By John Bonnoit de Maineduc, M.D. Bensley, London, 1785.

† Ibid.

‡ This Mrs. Pratt published certificates of these cures in a

However indisposed British practitioners may originally have been to listen to the advocates of Mesmer, when they perceived years afterwards the continental journals constantly reporting cases treated on principles of animal magnetism; and when, furthermore, they heard that some of the most eminent men of science* in Europe had become proselytes to the doctrine, it was reasonable to expect that some intelligent member of the profession would throw aside all prejudice, fairly investigate the reputed evidence, and communicate the result of his conviction to his medical brethren. This was done in a spirit of candour and veracity by the late Mr. Chenevix, which I believe never has been impugned. It is well known that the members of the medical profession, especially in England, are very scrupulous in conceding their belief publicly to any new science, or system of philosophy. The apprehension of ridicule, the policy of not outstepping the boundaries of faith prescribed by the prevailing opinions of society, are so generally felt, that few medical men of eminence will peril their reputation in advocating a doubtful, much less obnoxious, doctrine. It required, therefore, no ordinary degree of moral courage for any medical practitioner

pamphlet, entitled "A List of a few Cures performed by Mr. and Mrs. Loughborough, at Hammersmith-terrace, without Medicine." By a Lover of the Lamb of God. It was dedicated to the Archbishop of Canterbury, and is pervaded by a tone of fanaticism which could not fail to have been extremely offensive.

* In the year 1815, an hospital was established at Berlin, in which no medicines but Mesmerism and the prescriptions of lucid somnambulists were had recourse to. The eminent Hufeland, originally an unbeliever, was the principal physician of this hospital; and fifteen volumes containing the clinical details of the cases magnetically treated have been since published.

to throw down the gauntlet, and declare himself an out-and-out champion of the principles of animal magnetism. He, however, who has the temerity thus to stem the tide of popular prejudice, and openly advocate the cause of truth, will, in the end, receive his reward; if he live not to witness the hour of triumph, his memory will be yet crowned with honour.

In the history of animal magnetism there is, perhaps, no chapter which contains such strong and conclusive evidence in its favour as that which supplies us with details respecting the conversion of those who were originally unbelievers. After Mr. Chenevix had convinced himself of the truth of animal magnetism—which he terms Mesmerism—to avoid implicating the facts with the theory of the magnetic fluid, he announced the truth of its principles to the profession, and gave the following account of the circumstances which led him to that conviction. I quote his own emphatic words:—"Animal magnetism is true. In the whole domain of human acquirements, no art or science rests on experiments more numerous, more positive, or more easily determined. As this assertion is in direct contradiction to the vast majority of current prejudices, it is just to state the grounds upon which it is made. In former times, whenever animal magnetism was mentioned, I joined the tribe of scoffers; and so much was I convinced of its absurdity that, being at Rotterdam in 1797, I laughed to scorn a proposal made to me by an English resident there to witness some experiments in which he was then engaged. In 1803 and 1804, while travelling in Germany, I heard many very enlightened men of the universities talk of animal magnetism nearly with the same certainty as of mineral magnetism; but their credulity I set down to the account of German mysticity, and thought it not incongruous that the nation which took its philosophy from Kant, Tiehte, and Schelling, should believe that motions of the hands could, by the will of the mover, transmit an influence to the person acted upon, which should produce the wonder related of animal magnetism. I remained an unbeliever. In 1816 some persons of my acquaintance proposed to take me to the house of a lady in Paris, whose daughter was an artificial somnambulist, and in the terms of the art, lucid. I went to laugh; I came away convinced. To suspect anything like trick in the parties concerned was impossible;—they were of the highest respectability and distinction, and some of them I had known for many years. .... No sooner had the Abbé Faria begun to operate than the countenance of the young lady changed, and in two seconds she was fast asleep, having manifested symptoms which could not be counterfeited. The sitting lasted about two hours, and produced results, which though I still remained a sceptic on some of the most wonderful phenomena, entirely convinced me of the evidence of a Mesmeric influence, and of an extraordinary agency which one person can, by his will, exercise upon another. The Abbé Faria offered every means to dispel my remaining doubts; and gave me all necessary instructions to obtain total conviction from experiments of my own. I most zealously attended his labours, public and private, and derived complete satisfaction upon every point relating to Mesmerism; even upon those which appear supernatural. Many of the experiments I repeated, not only upon persons whom I met at his house, but upon others totally unacquainted with him and with his studies, and was ultimately compelled to adopt the absolute and unqualified conclusion announced above, 'Mesmerism is true.'"

Hence it will be observed that the belief of Mr. Chenevix in the truth of animal magnetism was founded on direct positive evidence;—on his own personal observation; and, more than all, on his own experience; for, as he subsequently states, nothing can be fairer than the proposition of the magnetisers, who say, "Come and see;—go and try." He did so, and what was the result? He informs us, that "of 164 patients, ninety-eight manifested undeniable effects; some in one minute, some not until the operation had been repeated several times. There was hardly an instance where disease existed that relief was not procured; and many of the patients offered phenomena as extraordinary as any recounted in Germany or France."........... "I was myself an unbeliever," he adds, "until I was undeceived by my own experiments; but had I sooner taken this plain and rational road to knowledge, instead of thinking all men mad who trusted to their eyes that told them truths which to me seemed more marvellous than all the other wonders of creation, I should many years since have possessed the conviction which I now enjoy, and not bewail, that in 1797, my presumptuous ignorance had shut, in my own face, the door of a science more directly interesting to man than all that chemistry and astronomy can teach. Nine-tenths who will read may laugh at this, as I did at my friend at Rotterdam; let them do so, but while they laugh, let them learn, and not, thirty years afterwards, have to lament that so short a remnant of life is left them to enjoy this new and most valuable secret of nature."* Acknowledg-

* Mesmerism, improperly denominated Animal Magnetism.

ments of this description from men of scientific eminence ought to have some weight. It is a description of evidence which cannot be repudiated. It were, indeed, preposterous to suppose that the voice of factious ignorance should set all human testimony at nought. He who has the temerity, however, to act even upon this outrageous principle, may, in the present case, satisfy himself by personal investigation of the truth of every fact I have asserted.

In 1833 the attention of men of science and literature in England was again called to animal magnetism, by Mr. Colquhoun's translation of the Report of the French Academy, to which was annexed a short essay, containing an abstract of evidence in proof of the occasional transference of the faculties, during somnambulism, from their usual and appropriate organs to the epigastrium, and other parts of the nervous system. Hitherto, as I have already stated, the old report of the Academy, in 1784, had been constantly, albeit ignorantly, referred to as evidence against the doctrine; but now the English public was put in possession of experimental details which remain unanswerable. The most obstinate sceptic could scarcely summon courage to challenge the concurrent testimony of the academicians,—men who were not the partisans of animal magnetism, but many of them avowedly unbelievers, and who, throughout their report, restrict their attention to the relation of the facts alone which fell under their observation. "It is curious, and by no means uninstructive," says Mr. Colquhoun, "to observe the different reception which these two Re-

By Richard Chenevix, Esq. F.R., and E.S.M., R.I.A., &c.—London Medical and Physical Journal, vol. vi. pp. 219, 227, &c. 1829.

ports respectively met with in the scientific world. The former, with all its numerous faults, imperfections, inconsistencies, and contradictions on its head, was at once almost universally hailed, by the professional physician and the philosopher, with the highest satisfaction and applause, as conclusive with regard to the reality, the merit, and the utility of an alleged important discovery which had begun to disturb the calmness of their scientific repose. The latter has been viewed with suspicion and distrust, and treated with censure, contumely, and ridicule, because it has opened up an obnoxious but highly interesting discussion, although this last committee, carefully avoiding the controversial example of their predecessors, have merely laid before their brethren the result of their own experiments and observations, without one word of argument, or a single allusion to theory. This affords one instance, among many, of the extreme reluctance which is felt by philosophers to allow their partial convictions to be unsettled by new lights, and of the great difficulty of procuring a favourable reception for doctrines which are objectionable only because they are deemed incompatible with preconceived notions."* In 1836, the Isis Revelata, being an inquiry into the origin, progress, and present state of animal magnetism, was published by the same author ;—the mass of information which it contains—the philosophical spirit in which the whole argument is conducted, could not fail to ensure for the author the respect and esteem of all who are interested in the cause of science. This work forms an epoch in the

* Colquhoun. Preface to the second edition of the Isis, vol. i. p. 23.

history of animal magnetism in England, analogous to that which was formed by the appearance of the Histoire Critique, by Deleuze, in France.

On my arrival in England, in 1837, I found that many persons who attended my demonstrations possessed a general knowledge of the principles of animal magnetism; but there being no permanent magnetic clinique in this city, the majority had never had an opportunity of personally observing the effects which are produced. This *desideratum* it was my intention to supply. At the North London hospital, in the presence of many scientific persons, professors of medicine, and students, I commenced my operations, and I feel myself fully warranted in stating that I succeeded in convincing many members of the profession of the reality of the magnetic influence.* Hence Dr. Elliotson, in one of his clinical lectures, delivered shortly afterwards, stated that he had already, in 1829, been convinced by Mr. Chenevix that there was something more than imagination in the effects produced; but that he had neither opportunity nor time to proceed with the investigation until my arrival in town. He then details the symp-

* An ably written pamphlet was published shortly after I had commenced my demonstrations, entitled "A Short Sketch of Animal Magnetism, intended to direct attention to the propriety of practically examining that question. By a Physician. Hatchard and Son, London." To the author of this brochure, and to some other gentlemen who are not personally known to me, I beg to return my acknowledgments for the unsolicited public testimony they have borne to the integrity with which my *séances* are conducted. The good opinion of conscientious men does more than compensate for the abuse and calumnies by which I have been assailed by certain critics, whose personal obscurity, doubtless, inspired them with the usual assassin-courage of anonymous slanderers.

toms which were exhibited by several patients magnetised by me at the hospital, and concludes with this declaration :—" Now, so far as the facts had gone, that is, those that had come under his own notice, he believed in what he should call Mesmerism, ... he was never ashamed to declare what he believed; he had little respect for authorities ; when he saw facts like those he had observed in the cases manipulated on by Baron Dupotet, he must believe them. The whole profession might laugh, but he must believe that there was a peculiar power which gave rise to the phenomena which he had observed, and that it was not sufficiently known or appreciated."* It is to me gratifying to add, that the magnetic mode of treatment has been, since this period, adopted in the wards of this hospital, and that several members of the profession have been induced to try it in private practice.†

Hence I am sanguine enough to believe that animal magnetism in England will, ere long, attain the same rank as a science which it already enjoys on the continent; and in the meantime, undisturbed by the ridicule of the ignorant or the abuse of the vulgar, I shall persevere in demonstrating its principles, fully assured that the innate force and power of truth will eventually triumph over the hostilities which may temporarily embarrass, but never permanently affect, its progression.

* Report of Dr. Elliotson's Clinical Lecture, in the Lancet for Saturday, September 9, 1837. Vol. ii. p. 866, *et seq.* Vide Appendix I. II. III.

† Vide Appendix IV. V.

CHAPTER XII.

EXPLICATION OF THE PHENOMENA OF ANIMAL MAGNETISM—TENDENCY OF THE DOCTRINE.

The phenomena of animal magnetism having been repeatedly demonstrated to the satisfaction even of the most incredulous, various speculative attempts have been made to explain them. The human mind is naturally disposed to theorise ; restless of advancement, it is apt to supply the void of real knowledge by phantasies of its own creation. Hence, in the early ages of the world, eclipses, meteors, every unusual appearance in nature, was readily accounted for by supposing it a special revelation from heaven to warn kings and kingdoms of their approaching destiny. This intuitive desire of satisfying our conviction by prematurely explaining, in some way or other, every fact which falls under our observation, has given rise in science to a host of absurd theories, many of which have seriously impeded the progress of human knowledge. Upon this principle we have seen that the ancients, the Egyptians, Greeks, and Romans, ascribed all the phenomena of somnambulism to direct inspiration from the gods ;—so also, in the middle ages, the learned and the unlearned regarded somnambulists so affected by the arts of sorcery and witchcraft, as demoniacally possessed;—and, with nearly as much sagacity, the commissioners of the French academy, in 1784, ascribed all the phenomena to three omnipotent causes—imagination, imitation, and manual contact (*attouchement.*) In thus prematurely hazarding imperfect explications of the magnetical phenomena, it is not to be concealed that many of the inveterate antagonists of the doctrine imagine that they strip the facts themselves, which, be it observed, they are constrained by the force of evidence to admit, of their specific identity as magnetical results, and reduce them to the level of ordinary effects arising from causes altogether independent of the magnetic influence. They strain their ingenuity to devise a theory for the express purpose of annulling the importance of the very facts which they are bound to acknowledge;—a motive which is certainly unworthy of any person who professes to be the advocate of truth. Imagination has been one of the most favourite of these hypotheses. It has solved more mysteries than the sphynx herself. Let us, however, examine *seriatim*, these theories; we shall find not one of them is at all adequate to explain the magnetical phenomena.

I. Imagination.—Without entering into any lengthened discussion on the powers and effects of the imagination, I shall briefly refer to certain facts already detailed, which must completely subvert this theory:—

1. It has been shewn that infants are very easily thrown into a magnetic state.

2. Persons, during their natural sleep, are very susceptible of the magnetic action; they soon, under its influence, become disturbed, and somnambulism and all the higher magnetical phenomena successively ensue.

3. It has been most distinctly proved that persons can be magnetised at a distance, through the intervention of screens, doors, and thick walls, when they are not aware even of the presence or intention of the magnetiser.

4. Animals, the organisation of whose nervous system is similar to ours, [the vertebrata] may also be magnetised.*

* From the Memoirs and Recollections of Count Ségur, the ambassador at the courts of Russia and Prussia, we learn that, about the year 1778, the doctrines of Mesmer excited a considerable sensation in Paris. "I was accompanied," observes the Count, "in this singular school by men whose names are not without some weight in the scale, since among them I am enabled to cite those of Count de Gébelin, Olavidès, d'Esprémének, de Jaucourt, de Chastellux, de Choiseul, Gouffier, de la Fayette, and a crowd of other persons distinguished in letters and the sciences, without mentioning medical men who were secretly concerned in the business, and who will no doubt give me credit for not writing them on my list as bashful Mesmerians, privately agreeing there was some truth in the discovery, but censuring it openly out of respect for the faculty." He then relates the following anecdote:— "The queen, one day, entered on this subject with me, and seemed to take a pleasure in repeating all the severe and amusing jokes which were circulated in such abundance at our expense. It was in vain that I wished to discuss the matter; she would not hear me, and merely said,—' How can you expect us to listen to your follies, when seven commissioners of the Academy of Sciences have declared that your magnetism is only the effect of a heated imagination?' Madam, replied I, rather hurt by the above remark, I respect this learned decree, but as veterinary surgeons have magnetised horses, and produced effects of which they attest the veracity, I could wish to be informed whether it was those horses that had too much imagination, or whether it was the learned doctors who were deficient in it. She laughed, and thus ended our conversation on animal magnetism."—Memoirs and Recollections of Count

Now, in not one of these cases can the imagination be assigned as a cause of the phenomena produced.

That the magnetic influence will take effect without the consent or sympathy of the party magnetised, was very clearly proved to the commission of the French Academy, by one of its own members, M. Itard, who, not believing that any such effects could be produced, subjected himself to the magnetic action. I operated, and the case, in their report, is thus related :—" When magnetised by M. Dupotet, on the 27th of October, 1827, he experienced a heaviness without sleep, a decided sensation of a peculiar nature ;—a setting on edge (*agacement*) in the nerves of the face, convulsive motions in the nostrils, in the muscles of the face and jaws, and a flow of saliva of a metallic taste ; a sensation analogous to that which he had experienced from galvanism." Two other cases were also submitted to the commission ; one, a child of twenty-eight months ; the other, a deaf and dumb boy, both of whom were magnetised. " The one," continues the report, " was not in a state capable of knowing it ; the other never had the slightest idea

Segur. From the French, 2 vols. London, 1826. Vol. ii. pp. 52, 53.

" Granting, for a moment," said Deslon, in 1780, " that Mesmer possesses no other secret than that of employing the imagination in the extensive production of the most salutary effects, will it not still be true that the invention is an extremely valuable one? For, in reality, if the physic of the imagination be more salutary than other kinds of medicine, what good reason can be alleged why the physic of the imagination should not be brought into general use?"—Observations on Animal Magnetism, pp. 47, 48.

of magnetism; both, however, were sensible of its influence, and it is impossible, in either case, to attribute this sensibility to the imagination; still less is it attributable to this principle in the case of M. Itard. It is not over men of our years, and like us, always on their guard against mental error and sensible delusion, that the imagination, such as we view it, has any sway. At this period of life it is enlightened by reason, and disengaged from those illusions by which young persons are so easily seduced. At this age we stand upon our guard, and *distrust* rather than *confidence* presides over the different operations of our minds. These circumstances were all happily united in one colleague, and the Academy knows him too well not to admit that he really experienced what he declares that he felt."*

The magnetisation of persons during sleep, or in a state of insensibility, proved also, to the satisfaction of Cuvier, whose authority may perhaps be as highly esteemed as the report of any Academy in Europe, that the magnetic effects cannot be explained by any such hypothesis. In the second volume of his Comparative Anatomy, page 117, he observes, "It must be confessed that in all experiments, the object of which is to determine the action which the nervous system of one person may have on the nervous system of another, it is difficult to distinguish the effects of the imagination of the person acted upon from the physical effects produced by the person who acts. Yet the effects produced upon persons who before the operation was begun were in a state of insensi-

* Report of the French Academy, Eng. Translation, Isis Revelata, vol ii. pp. 216, 217.

bility,—those which have taken place upon other persons, after the operation itself had reduced them to that state, and also the effects produced upon animals,—no longer permit it to be doubted that the proximity of two animated bodies in a certain position, and with the help of certain motions, do produce a real effect wholly independent of the imagination of either. It is also evident that these effects are entirely owing to a communication which takes place between the nervous systems of the two parties."

II. Imitation.—This cause might be expected to operate to a certain extent, were the magnetic process always conducted in an assembly where many persons were at the same time being magnetised ; but in silence and privacy, when the magnetiser is alone with his patient, the operation is attended with the most striking effects. On such occasions, when there is no person present to imitate, the same effects are produced. How is it, too, that ignorant persons, the peasantry, for example, in Germany, Switzerland, or Holland, when magnetised, exhibit the same phenomena, albeit they never heard, and could have no idea, of the effects which would attend the operation ? How is it that so constant an uniformity in the results produced by magnetism prevails in all parts of the world ? Surely this very fact proves that the effects of animal magnetism depend on certain fixed principles, which, like other of the laws of nature, are similar in all latitudes.

III. Manual contact, or attouchment. — This theory is at once subverted by the fact, that the best magnetisers produce all the magnetical effects without the manipulation *per contactum*. It is a process which is now seldom or never used.

IV. Animal heat.—This is absolutely an absurd

theory. Individuals constantly sleep together and communicate animal heat to each other, but no magnetical effects are the consequence. Besides, the magnetic operation is often performed at such a distance from the patient, that no communication of heat even by radiation could possibly be felt.

In addition to these theories, which do not carry with them even a colouring of plausibility, many lay considerable stress on the effects which the monotonous manipulations of the magnetiser must have on the patient. It is true that monotonous actions and sounds,—a rocking cradle, a swing, an old lullaby, the murmurs of the sea,—anything which by constant repetition fatigues the attention, will induce sleep; but let it be observed, that the sleep which is so produced is a perfectly natural sleep, there is neither insensibility, somniloquism, clairvoyance, or any of the phenomena which characterise the magnetic state. No monotonous motions of the hands merely before any person will give rise to these effects; they depend, as I have already explained, on the will of the magnetiser, and the monotonous movements of the hands may, in many cases, be altogether dispensed with; they are not necessarily a part of the operation. Lastly; a theory has been lately proposed by Dr. Sigmond, who, having operated successfully in many cases, is of opinion that the "art consists in obliging the individual again to inspire by the nostrils the carbon he has already expired, whilst the currents of air caused by the extended fingers produce some effect on the facial nerves."* It is true enough, as

* Dr. Sigmond's Communication to the Lancet, *loc. cit.*, vol. i. p. 387.

travellers amuse themselves by proving at the Grotto del Cane that the inspiration of carbonic oxide gas induces stupor; but Dr. Sigmond will, I doubt not, be able himself to draw a diagnosis between the stupor arising from the inhalation of narcotic gases and the insensibility of the magnetic state; besides, it may reasonably be doubted whether the passing the fingers up and down before the nostrils, or even wrapping the mouth round with a cotton comforter, which might occasion a person to inhale repeatedly the breath he had expired, would lead to the accumulation of a sufficient quantity of this gas in the lungs to produce even stupor: this again is an untenable hypothesis. Hence not one of these theories affords any satisfactory explication of the magnetical phenomena which are developed.

I now, therefore, proceed to detail the theory which the magnetisers of the present day have adopted; but let it be remembered that whether accepted by the reader or not, the facts on which it is founded still remain unimpeached. In so doing, I give, in particular, the views of Rostan, an eminent pathologist, who, like other men of science, could not withhold his conviction as to the truth of animal magnetism, when he had witnessed personally its effects. He states, in the following terms, the creed of the magnetisers:—"It is our opinion that the whole of these phenomena belong to the nervous system, all the functions of which are not as yet known; that it is to a modification—to an extension of this system, and of its properties, that the effects we are now considering must be ascribed. In the present state of science, everything leads us to consider the brain as an organ secreting a peculiar substance, the principal property of which is to transmit and to receive volition and sensation. This substance, whatever it may be, appears to circulate in the nerves, some of which are appropriated to motion (to volition); these proceed from the encephalon, or its dependencies, to the extremities: the others belong to sensation, and these proceed from the base of the encephalon. The first are active, the second passive.

"We may consider the following propositions as established:—When we wish to move any limb, our brain sends to the muscles destined to execute the movement a certain quantity of the nervous agent, which produces contraction. This transmission is effected by means of nerves shewn by anatomy; and if we cut them or tie them, it becomes impossible to execute the motion; paralysis ensues.* The same phenomenon takes place with the nerves of sensation; if they are destroyed, sensibility is annihilated in the part to which they proceed. These facts, known from time immemorial, are incontrovertible, and generally admitted. They gave rise to the opinions, that the function of innervation was an actual circulation:—that there were nervous vessels *afferent*, those of the will; and *efferent*, those of sensation. The labours of M. Bogros proved experimentally that which had been previously arrived at by reasoning. But what is the nature of this agent? The researches of MM. Prevost and Dumas lead us to believe that the

* The circumstance of the parts *below* the division of the nerve being paralysed, and not those *above* it, would appear to prove the circulation of a nervous fluid from the centre to the circumference of the body, as clearly as the experiment of the ligature by Harvey proved the circulation of the blood downwards in the arteries and upwards in the veins.

strongest analogy exists between this agent and the electric fluid. These physiologists demonstrated that muscular contraction was the result of a real electric shock. M. Beclard affirmed, that having laid bare and cut through a nerve of considerable volume in a living animal, he frequently caused the pole of the magnetic needle to deviate, by bringing this nerve in contact with the needle. It is known by all persons, that galvanism, substituted for the nervous influence, forces the muscles submitted to its action to contract. Galvani and Volta saw and proved the existence of a peculiar fluid which, at a later period, has been recognised as identical with electricity.

"It is known, also, that certain animals possess the singular property of secreting, by means of their peculiar organisation, a large quantity of the electric fluid, with which they give violent shocks ; so violent sometimes as to occasion the death of other fish, or even of human beings, within a certain distance from them. The *Gymnotus electricus*, the *Silurus electricus*, the *Tetraodon*, the *Torpedo unimaculata marmorata*, and many others, possess this faculty. The quantity of their electric fluid has been appreciated by means of very delicate electroscopes and electrometers. It has been ascertained that this fluid is secreted by the brain of these animals, since by removing it, or the nerves which proceed to it, the electric effects are annihilated—a result which does not take place when the organs of circulation are removed.

"Thus it is satisfactorily demonstrated, that in some animals the brain secretes a portion of the electric fluid; that muscular contraction may be produced by electric excitation, &c.—a consideration which gives us strong reason to presume that the nervous agent is either electric fluid, or a fluid very analogous to it. We shall pass over in silence the proofs that might be deduced from acupuncturation and perkinism,* &c. But this agent does not confine itself within the muscles and the skin; it throws itself off with a certain degree of force, and thus forms a real nervous atmosphere—a sphere of activity absolutely similar to that which surrounds electrified bodies. This opinion is also that of some of the first physiologists of our time—Reil, Authenrieth, and M. Humboldt; and being admitted, all the phenomena of magnetism appear to be susceptible of a plausible explanation.

"The nervous and active atmosphere of the magnetiser, increased, no doubt, by the impulse of his volition, mixes, and is brought into contact, with the nervous and passive atmosphere of the magnetised person, which latter it augments to such a degree, that in some cases there seems to be an actual saturation of the nervous system, capable, when in excess, of putting itself in equilibrium with surrounding bodies; and there is no way of explaining by any other hypothesis the *secousses* sometimes experienced by the patients. The nervous system of the magnetised person being thus influenced, and undergoing modifications in proportion to his peculiar sensibility, would suffice

* "I confess I have always been much disposed to think that electricity, modified by the vital action, is the invisible agent which incessantly traverses the nervous system."—Cabanis.

"Organised beings, and especially the human body, composed of the assemblage of a great number of heterogeneous substances in contact with each other, present to us a real and complicated electrical apparatus, in which the principle of which the nerves are the conductors, appears to act in a manner analogous to that of electricity."—Physiologie de Richerand, t. ii. p. 263.

to explain all the perturbations observable, and would perfectly account for the communication of the desires, the will, and even thoughts, of the magnetiser. These desires, this will, being actions of the brain, it transmits them, by means of the nerves, as far as the periphery of the body, and beyond it. In this sketch the full mechanism of the production of magnetic effects has probably not been unveiled. This hypothesis, however, which does not depart much from the physiological and physical facts generally admitted, would explain, to a certain extent, the greater part of the innumerable effects produced by animal magnetism, and may, perhaps, assist in revealing some of the most astonishing mysteries of animal life."

There are many persons who, I am aware, object in unmeasured terms to this theory of the magnetic fluid; but there are certain facts which I have related in this volume, which are inexplicable on any other hypothesis.

1. It is well known that if, during the magnetic operation, the magnetiser, instead of magnetising from above downwards, makes a counter-current; if he intentionally reverse the manipulation, directing the fluid suddenly from below, upwards, the patient will become much disturbed, and muscular spasms, or even convulsions, will be produced.

2. It has been proved that persons in proximity with those who are being magnetised not unfrequently receive the magnetic influence indirectly, and without any intention or consent on the part of the magnetiser. This can only arise from their being incidentally placed within the sphere of a material effluence; for here, on the part of the magnetiser, no psychical cause is in operation.

3. It is experimentally proved that the fluid is propelled from the body in right lines, and that some few objects can altogether intercept its progress.

Lastly. It is clearly established that the magnetic fluid can be accumulated in certain inanimate objects, which, when grasped in the hand, communicate a shock somewhat similar to that which occurs under similar circumstances in electrical experiments.

The above facts lead me, therefore, to adopt the theory of a magnetic fluid being transmitted from the nervous system of the magnetiser to that of the person operated upon; but it is not to be forgotten that great mental energy, sustained concentration of the will, is necessary to direct and control its influence. He then, who is magnetised, passes into a state of complete physical insensibility, during which he awakens, as it were, within himself, and enters into a new mode of existence and relation with the external world, for all his perceptions are now exquisitely fine, and independent of the instrumentality of mere corporeal organisation. And if any person should ask what is the moral tendency of the doctrine of animal magnetism, I should answer, that it obviously tends to establish the spiritual ascendancy of man over those material conditions which, in his ordinary state of being, fatally restrict the apprehensions, capacities, and comprehensions of the soul; and this very manifestation of its existence, partially divested of the grosser elements in which it is temporarily obscured, affords a precursory evidence of a future state of being, which belief in itself cannot fail to suggest those principles of self-government and moral conduct which can alone promote the real welfare and happiness of society.

In the ordinary routine of life, persons act magnetically on each other, when perhaps they are least aware of it; thus, he who would obtain the esteem of another, must mentally exert all his energies to establish a reciprocity of feeling; he must act and produce an effect on the mind of the person he regards; and if he succeed, the affections of both subsequently become commingled, so that a perfect intellectual unity is induced; and hence the novelist did not exaggerate truth when he described two such beings as moving with "one soul in a divided body." In various conditions of life, we may observe the magnetical principles are brought into operation. The orator, endeavouring to move the feelings of his hearers, rouses up and concentrates all his nervous energies to effect his object, and led gradually away by the spirit of his persuasive eloquence, he passes into a state of excitement almost identical with magnetic extase, during which the divinest language flows almost unconsciously from his lips, for he is sensible of no mental effort in arranging the consecutive order of his thoughts, and the construction of his sentences; his ideas crowd upon him unsought, and are evolved with preternatural rapidity, so that he appears like one who is inspired, until his enthusiasm is over, and then, on returning to his ordinary state of being, he finds himself unable to reconstruct with the same harmony, and beauty, and power of language, the very oration he has just delivered.* Hence the ancients drew a marked distinction between eloquence and

* The account which Lord Brougham gives, in his Paley's Theology, of an orator busily constructing one sentence in his mind while he is uttering another, cannot be analytically made out; it will be observed, too, that his theory is at variance with the unity of consciousness.—Vide Paley's Theology. By Lord Brougham.

mere verbal oratory. "Eloquence, indeed," says Pliny the younger, "is the privilege of very few, nay, if we will believe Marcus Antonius, of none; but that faculty, which Candidus calls loquacity, is common to numbers, and the talent which generally attends impudence."* Again, the physician also, to be successful, must act upon magnetical principles; he must constantly maintain a mental power or ascendancy over the mind and nervous system of his patient, in order that he may possess his entire confidence; and if this relation, or truly magnetic rapport, be not established, all his skill will prove unavailing. It may, indeed, with truth be affirmed, that the psychical influence of the physician over his patients, effects more good in many cases than the physical remedies which he prescribes. In domestic life, the magnetical sympathies by which individuals are associated together, and their affections consolidated, may often be strikingly observed; but many, perhaps, are not aware that the proximity of two persons to each other, so intimately commingles the nervous atmosphere by which each is surrounded, that there is an actual transference of vitality from the body of the one into the body of the other. This is no nursery dream—no gossiping fiction—but a fact which is well known to physicians. Hence Dr. James Copland, in his learned and very admirable Dictionary of Practical Medicine, observes, that "a not uncommon cause of depressed vital power is the young sleeping with the aged. This fact, however explained, has been long remarked, and is well known to every unprejudiced observer. But it has been most unaccountably overlooked in medicine. I

* Letters of Pliny, the Consul, with occasional Remarks. By William Melmoth, Esq. 2 vols. London, 1748. Vol. i. p. 298.

have on several occasions met with a counterpart of the following case :—I was, a few years since, consulted about a pale, sickly, and thin boy, of about four or five years of age. He appeared to have no specific ailment, but there was a slow and remarkable decline of flesh and strength, and of the energy of the functions—what his mother very aptly termed, a gradual blight. After inquiry into the history of the case, it came out that he had been a very robust and plethoric child up to his third year, when his grandmother, a very aged person, took him to sleep with her; that he soon afterwards lost his good looks, and that he had continued to decline progressively ever since, notwithstanding medical treatment. I directed him to sleep apart fom his aged parent, and prescribed gentle tonics, change of air, &c. The recovery was rapid. But it is not in children only that debility is induced by this mode of abstracting vital power. Young females married to very old men suffer in a similar manner, although seldom to so great an extent; and instances have come to my knowledge where they have suspected the cause of their debilitated state. These facts are often well known to the aged themselves, who consider the indulgence favourable to longevity, and thereby often illustrate the selfishness which in some persons increases with their years."*

This transference of vitality is thus well marked in cases of extreme disparity of years between the parties approximated, as when the young are placed in contact with the aged; but the same transference, doubtless, will take place between persons of any age, although, where the vital principle of the two persons

* Dictionary of Practical Medicine. By James Copland, M.D., F.R.S., &c. Art. Debility, vol. i. p. 475.

exists nearly in an equilibrium, the effects will be less perceptible. Here, also, I would remark, that precisely on the same principle the mother acts magnetically on her child; she concentrates her thoughts and feelings on the object of her solicitude, and infuses into its yet unconscious bosom the elements of her own physical and moral constitution, so that, by this transference, the seeds of good or of evil are sown in the tenderest years of infancy, and in this sense is to be understood the scriptural phrase, that a tree shall be known by its fruits; not that the blind physical organisation can lead of itself to any such consequence, but that the spirit, which is the life even from within the trunk, shall permeate the remotest branches, and either give beauty to the flower and goodness to the fruit, or impregnate both with poison more deadly than the blight of the fatal Upas-tree, which is reported to kill all that inhale the atmosphere around it. The principles of animal magnetism thus lead us to perceive relations between physical and moral conditions of humanity, which were before a perplexing mystery; they throw light on a variety of facts hitherto deemed inexplicable, if not incredible, in the early history of mankind; they place us in possession of a power whereby we can alleviate suffering and restore health to the afflicted; they lead us to entertain also the spirit of a philosophy which is of the most cheering description, annihilating as they do all those dark attributes of materialism which have so long thrown a gloom over the paths of Science. But before these relations can be perceived, or these consequences deduced, it is necessary that the facts themselves should be clearly, distinctly, and most unequivocally established. And how is this to be effected? By positive evidence; by ocular demonstration. Let me, therefore, once more entreat men of science and literature to witness the facts, and scrutinise them as narrowly as possible; and I feel assured that many who may now lay this volume down with a smile of incredulity, will be induced to rescind their scepticism, and allow that I have "given faithful record and fair inference," and, in the words of my epigraph, "reared a superstructure on just grounds."

APPENDIX,

CONTAINING

REPORTS OF BRITISH PRACTITIONERS IN FAVOUR OF ANIMAL MAGNETISM.

"The mere account of foreign cases is not enough. The sober good sense of British practitioners requires that what was done at Paris, or Vienna, should be capable of repetition in London."

Medical Quarterly Review. London, 1835.

APPENDIX.

I.

REPORT OF CASES TREATED MAGNETICALLY AT THE NORTH LONDON HOSPITAL, BY DR. ELLIOTSON.

(Extracted from the *Lancet*, Saturday, September 9th, 1837.)

THE first case was one of epilepsy, occurring in a girl sixteen years of age, a housemaid of diminutive stature. She had been subject for twelve months before her admission, on the 4th of April, 1837, to attacks of epilepsy, which occurred about once a week, or oftener. At twelve years of age she had a fall, by which she was stunned: this was worthy of remembrance. She was subject to almost constant headaches in the morning and evening; they also came on a short time previous to the occurrence of a fit, and sometimes shooting pains across the occiput preceded the fit for a few days. She also experienced before the fit came on a sensation of coldness, which ran up the spine, and was attended with numbness, which, when it reached the head, produced the feeling of her being stunned, and then she lost consciousness. The fit was characterised by convulsions, chiefly of the face and trunk; the extremities were in a rigid state, the hands clenched, the face variously contorted, the eyes rolled. This state was not followed by coma, but with a restlessness and sleeplessness, and a severe throbbing pain of the head, which generally continued for three or four hours. She had been cupped on the shoulders last January, and had leeches applied to the temples, and took some aperient medicine. She had a fit last night; the fits usually come on in bed during the state intermediate between sleeping and waking. Complained on her admission of headache, and sickness after food; tongue clean, appetite bad, bowels regular.

The peculiarity in this case consisted in the fact of the patient, as soon as the convulsions of the epileptic seizure were over, subsiding into a restless, fidgetty state, which lasted three or four hours, instead of falling into the state of coma, which is usual in this affection. In the treatment of the case, at first, as the pulse was not full, it was thought there was no necessity for blood-letting, and low diet was considered sufficiently active antiphlogistic treatment; and on the 4th of April quarter-grain doses of nitrate of silver were commenced, and given three times a-day: the dose was increased by a quarter of a grain gradually, until she reached two-grain doses three times a-day. On the 9th, the report stated that since the last fit, which occurred on the 6th, she had suffered from continual pain in the temples, and, indeed, all over the head, and it was found necessary to take blood, which was buffed and cupped. She was now subject to similar attacks, and it was found necessary frequently to bleed her; the pulse was hard and full. On the 16th of May she was bled to eight ounces; on the 20th, to eight ounces; and on the 23rd ten ounces were taken away. The nitrate of silver was being increased all this time gradually. She took this medicine for six weeks from its first commencement; he (Dr. E.)

never gave it during a longer period, for fear of discolouring the skin. On the 27th of May, the fits being as frequent as before, though the nitrate of silver was given in two-grain doses, that medicine was discontinued, and the cuprum ammoniatum, in quarter of a grain doses, was commenced, and given three times a-day, the dose being increased by a quarter of a grain twice a week. On the 5th of June she was taking one-grain doses of the medicine; this produced sickness, though, as has been stated, she bore two-grain doses of the nitrate of silver. As the copper produced this nausea, the dose was diminished to three-quarters of a grain; and as this now, also, produced sickness, on the 10th she took one minim of creosote with each dose of it, and this effectually prevented the nausea, and the copper was again increased to grain doses. On the 16th, the report states, that there has been no nausea and no fit. On the 17th, however, the nausea returned, and the creosote was increased to two minim-doses, with the effect of checking the sickness, and she bore the copper well. On the 20th, the dose was increased to one grain and a quarter. On the 24th, there had been no fit, and there was no nausea. Soon after she began the copper, which had not produced any decided effects, another agent was called into action in the cure, and any alteration in the state of the case he (Dr. E.) considered was to be ascribed to this.

Under this treatment, which he should shortly speak of more fully, the fits ceased altogether, and instead of the patient having convulsions as she used to have, she was now seized with fits of somnambulism, or, as it had been proposed to call the state, somno-vigilium—a much more appropriate term, and one which expressed the condition better, the patient being both asleep and awake, and walking not being necessary for its existence, that symptom being present in only one species of the affection. All at once she would become perfectly insensible, but her eyes would remain wide open, but perfectly insensible to the effect of light: pulling her hair produced no impression on her. Her sense of hearing was lost to all ordinary sound. Though her eyes were open, yet she was perfectly blind; when you dashed the fingers suddenly towards her eyes, there was no winking. Her tongue was not tried, but it is probable her sense of taste was gone. She had never, in her own recollection, had the sense of smell. Yet, though now totally without external sensation, she was constantly talking, and talked very sensibly, and wittily too; but from the great variety of topics her conversation embraced, it amounted to rambling. She displayed, also, a great spirit of mimicry. This state had been commonly called somnambulism; somno-vigilium, as he (Dr. E.) had said, was a better term, for this patient could neither stand nor walk, but laid in bed during the attacks. This state would cease as a fit of epilepsy or hysteria ceases. She would become suddenly still, look wild for an instant, rub her eyes, be sensible of everything around her, and resume her natural character, which was that of a quiet, modest girl. These attacks were remarkable, both from the sudden manner in which they came on, and the equally sudden manner in which they went off. As she could not be awakened by external impressions made on her, Dr. Elliotson determined on trying what would be the effect of producing a strong internal sensation in her, and endeavoured to get her to believe she was likely to fall to the ground. For this purpose she was lifted from the bed, and it was found that she could not stand; she was then supported under the arms, and at first she felt the ground slightly, then afterwards gradually got a proper feeling in her feet, then by degrees she began to step out, and, with assistance, at last walked firmly. She suddenly awoke during the walk, ceased to speak, appeared lost for a moment, then moved her head, seemed astonished, and awoke, and walked to bed well. She was shortly after laid on her bed for repose, and no sooner was this done than she fell again into the same state. In a moment she again became senseless; her eyes, again, though wide awake, lost their sight entirely; she talked again in the same rambling way, and was again as rude in her mimicry as before. She was again taken up, and walked about, and was instantly restored, and remained in her natural state for a week. At the end of a few days the same kind of attacks again came on. Nothing had made so great an impression on him (Dr. E.) as to see the brain, as instanced in this case, completely senseless, and yet the mind wandering and active, in two or three moments the patient losing all external sensation, which was suddenly again restored, again lost, and again restored. Some similar symptoms were observed in a common epileptic fit, such as the person in a moment becoming perfectly senseless, and as suddenly becoming sensible, but the intellect, as in this case, was not active during the state of external insensibility. After a few weeks the character of these attacks completely changed; she still wandered in her talk, displayed the same spirit of mimicry, sang, whistled, danced, was rude, noisy, laughing, or miserable, by turns, but she also retained in these attacks all her external senses; she saw, heard, and walked well, and the power over the voluntary muscles was entire; she was now in a state of what was called ecstatic delirium; the attacks coming on suddenly, and as suddenly ceasing, no symptoms of phrenitis being present, the attacks lasting a few hours, and coming and going like attacks of epilepsy. There you might see her in the ward singing, dancing, and mimicing every individual; sometimes swearing, sometimes rather affectionate; then the attack going off as suddenly as it came on; she seemed lost for a moment, then began to smile, look pleased, and at once became herself again.

Now, they (the students) were aware, that when affections of this nature attacked persons while asleep, and they were partially awakened, the state they were then in was called somnambulism; but when the attacks came on while the persons were awake, and they became half asleep, the result was ecstacy. The one term he had employed, that of "sleep-waking," applied to both these states. Sleep-waking was much more extraordinary in some persons than in others. This patient recollected in one paroxysm what occurred in former ones, but when out of the attacks she forgot all that took place in them. She had a memory of circumstances which happened in the fits during the presence of the fits only; but she also remembered in the fits those things which occurred out of them. The eyes of this patient converged towards the nose, one rather more so than the other; he (Dr. E.) wished to know if she saw double; she appeared to see everybody; but her vision was much disturbed, for she never winked when the fingers were pushed suddenly towards the eyes, and when one finger was held up she said there were two; when two, that there were four; then she often said that people's "eyes were turning round," &c. This double vision was a common occurrence when one eye converged more than another, and arose from the adductor muscle on one side being drawn aside. When a watch was held to her, she could not tell the minute hand from the other, and it all appeared confused to her, though she seemed to the looker-on to have perfect vision. This symptom has been remarked in other similar cases, as well as the effect also of light on the iris, and pushing the fingers against the eyes, which produced no winking.

II.

FURTHER REPORT OF CASES TREATED MAGNETICALLY AT THE NORTH LONDON HOSPITAL.

(Extracted from the *Morning Post*, March 2nd, 1838.)

THIS singular subject is daily attracting more attention, and the contest respecting the reality or deceptiveness of the resulting phenomena is becoming more and more interesting. Our readers may range themselves on either side, as they please (or rather, as they must, for conviction is not a matter of choice.) We shall not attempt to do violence to their opinions, but shall content ourselves with laying before them the result of a patient attendance, and, we trust, impartial observation, on several occasions, within the last three weeks, in the wards and lecture-room of the North London Hospital, Gower-street. We trust that we have taken time enough to allow our first impressions and prejudices to subside, and may now safely recommend our readers to go and judge for themselves.

We there saw Dr. Elliotson and his assistant, Mr. Wood, in the course of their routine of visit, in the midst of a crowd of pupils and visitors, produce very extraordinary effects on two female patients, Elizabeth O'Key and Hannah Hunter, apparently with no geater mental and bodily exertion than what would attend the act of steadily holding forth the hand towards the forehead of the patient for a few minutes, varying from two to twelve, till the effect is produced. This treatment is, it appears, no longer looked upon as a matter of experiment or curiosity in the hospital, but as a portion of a regular (though very peculiar) medical course, which has been found in these two cases to be productive of decidedly good results. The patient to be thus acted on is seated in a chair—the doctor, or his assistant, silently extends one hand open and the palm downward, towards (but not touching) her eyebrows—and in a few minutes she breathes heavily, and at once drops off into a state of insensibility, which differs from sleep in some very extraordinary respects. Both patients grow cold at the extremities; their pulse becomes so languid as to be scarcely felt; and they cannot be aroused by any of the ordinary means which never fail to disturb the slumber of the soundest sleepers. The extent to which this has been carried ineffectually will be best seen in the medical notes of the treatment of these cases, which we subjoin. These we obtained from one of the pupils, who took an especial interest in these proceedings, and will doubtless be more acceptable to medical readers than our own unprofessional observations.

The most interesting case is that of the girl O'Key. Each time that we beheld her magnetised, or Mesmerised, she fell at first (for a minute or two) into the insensible state, and then stood up, exhibiting a new condition of existence, in which the ideas, language, and feelings, appeared thrown back into artless, ignorant, puerile childhood. The dialect she then gabbles volubly is scarcely English, thickly interspersed with "what" or "whatten," and "such" or "suchen," which seem, by the emphasis she imparts to them, to hold most important places in her vocabulary as expletives and superlatives. In her ordinary state she appears an intelligent gentle little girl, without any peculiar defect in her language, very pale (the effect of repeated bleedings, to relieve continual headaches), and is said to be, notwithstanding her sufferings in this respect, possessed of an excellent temper and affectionate disposition. Under the operation of magnetism, all her good sense appears lost, but her good feelings remain in full force, and she evinces deep sympathy with some one of the pupils present, in whom she takes a particular interest. * * * * * * * * * * * Her memory in the trance appears remarkably tenacious on some points—a perfect blank on others. When she first awakes into her magnetic state (if we may be allowed to use the expression), out of the preliminary minute of slumber, she seems to enjoy at once a revival of her previous magnetic-dreaming existence, inquires affectionately for Dr. Elliotson or some friends who has been lost to her mind's eye during her ordinary waking intervals, and then begins to chatter away her gipsy *patois*, free from headache, untroubled by prudential considerations, unconscious of any operations, however painful, which may be performed on her—tells comical stories, mimics the crooked-mouthed family, who could not blow out the little farthing rushlight, and whistles very prettily. Her senses meanwhile appear but half awake—her fingers at times lose their perception altogether, and her eyes the power of adjusting the optic axes, so that one finger held up appears to her as two. Her appearance as she sits, as pale and almost as still as a corpse, is strangely awful. She whistles, to oblige Dr. Elliotson—an incredulous bystander presses his finger on her lips—she does not appear conscious of the nature of the interruption, but when asked to continue, replies, in childish surprise, "It can't." This state of magnetic semi-existence will continue—we know not how long; she has continued in it for twelve days at a time, and when awakened to real life, forgets all that has occurred in the magnetic one! Can this be deception? We have conversed with the poor child in her ordinary state, as she sat by the fire in her ward, suffering from the headache, which persecutes her almost continuously when not under the soothing influence of the magnetic operation, and we confess we never beheld anybody less likely to prove an impostor. We have seen Professor Faraday exerting his acute and sagacious powers for an hour together in the endeavour to detect some physical discrepancy in her performance, or elicit some blush of mental confusion by his *naïve* and startling remarks. But there was nothing which could be detected, and the Professor candidly acknowledged that the matter was beyond his philosophy to unravel! He was, however, after a brief lesson in "manipulation," at the hands of Dr. Elliotson, able on the instant to restore both patients to their states of ordinary consciousness. The requisite movement is performed by drawing the thumbs (or a finger of each hand) from the top of the nose along the eyebrows, outwards to the temples, quickly and steadily, two or three times.

We shall now subjoin the professional notes bearing on the cases of these two girls, only observing that the medical student who drew them up employs the word "Mesmerism" to express the operation or influence, which is, in ordinary, known by the name we have employed as most familiar to the majority of readers,—viz., "animal magnetism."

NOTES BY A PUPIL.

"Elizabeth O'Key, aged seventeen, was admitted last April, labouring under epilepsy; the fits came on every few days, scarcely at longer intervals than a week. Baron Dupotet commenced practising Mesmerism in her case last summer. At first a deep sleep was produced, and during the continuance of this process, the fits came on at longer intervals. After a time, a different effect was produced, and fits of delirium, in which she lost entirely sensation, came on spontaneously at intervals, and were always removed by Mesmerism, but never caused without this process. Latterly, it has been ascertained that this delirium might be produced artificially by Mesmerism, and that it might be done without the patient being aware that the process was going on. In short, the delirium may be induced and removed without her knowledge. It is not necessary to the production of this effect, that any movements should be made with the hand, or even that the hand should be used at all. Any person may produce this effect, and also remove it. In this delirium she does not remember what takes place in her natural state, nor in her natural state does she remember what takes place in the delirium; but she distinctly remembers from one fit to another, though she has no idea of time, saying that a circumstance which occurred a month ago happened within a minute. She can recognise persons; but is not able to see the absurdity of a man being her mother! To describe all the peculiarities of her case would be almost impossible; it must be seen to be understood. A few days since it was found necessary to place a seton in the back of her neck. The insensibility was produced in a few minutes in the ordinary manner; and while she was talking, and in the middle of a sentence, the seton was thrust through at the back of her neck. She took no notice whatever of it, finished the sentence, continued talking, and evinced not the slightest pain, nor could the least trace of suffering be discerned in her countenance. The wound was dressed, her clothes re-adjusted, and she was then restored to her natural state in a few minutes, perfectly unconscious of having been the subject of any operation, as was evinced by her saying, some time afterwards, that she thought somebody had been pinching her neck, for it felt very sore, and was perfectly astonished when told that a seton had been introduced. Surely this circumstance alone is sufficient to justify an inquiry into Mesmerism. If by this means painful operations may be performed without the knowledge of the patient, why do surgeons hesitate for one moment to put its efficacy more extensively to the test?

"What further advantage medicine may hope to derive from the more extensive adoption of Mesmerism remains to be proved; time alone can solve the question. It is now upwards of three months since she had the last fit."

"Hannah Hunter, aged thirteen, was admitted the beginning of last month, having lost the use of her legs entirely, and being unable even to sit up in a chair. The treatment adopted was tonics, cold shower-baths, and electricity. After these remedies had been persevered in for about three weeks, she so far improved as to be able to sit up; but was still unable to stand alone. On the two first occasions of being Mesmerised, drowsiness only was produced; the third time the peculiar sleep caused by this process was induced and removed at pleasure. After this had been continued for a few days, all the other means of treatment were omitted, the sleep was induced every day, and allowed to continue an hour or two. For a few days there was no apparent change; but after that she began to improve, her strength increased in proportion as the paralysis subsided, and she is now able to stand very well, and even to walk with very trifling assistance."*

* Let the reader compare this case with the facts stated at pages 188, 192, *et seq.* In these experiments at the North London Hospital, it will be observed that all the physical phenomena of Animal Magnetism were produced in a very marked manner; so also were all the psychical phenomena, with the exception of clairvoyance and prevision, which only occur in the very highest degree of the magnetic state; these, I have no doubt, will in some future cases be developed, if the magnetic treatment be persevered in.

III.

FURTHER REPORT OF CASES TREATED MAGNETICALLY AT THE NORTH LONDON HOSPITAL.

(Extracted from the *Globe*, Tuesday, April 17th, 1838.)

We witnessed, yesterday, at the North London Hospital, some of the experiments of Dr. Elliotson in this so-called science. Without pretending to determine the value of these operations, or the nature of the conclusions to which they lead, we may describe, with as much accuracy as we can, what we actually beheld. So much, at least, is due to any individual who submits his professions to the test of publicity, and invites and encourages open investigation.

The first persons submitted to the operation were two boys of the same age, of apparently similar physical organization, affected each with the same disease, and both of whom had entered the hospital at the same time. In the countenances of the two a marked difference was however perceptible, and the most casual spectator might have at once decided that one was far more likely than the other to be affected by anything that appealed to the imagination, or was to work upon the nervous system. Accordingly, it was proved, that while the one was easily and rapidly subjected to the influence of the experimenter, the other remained absolutely unmoved. Dr. Elliotson took his station behind both, and commenced a conversation with the spectators in front. During the course of the conversation, without anything that might inform the patients of his movements, he commenced his magnetising operation by pointing the fingers at the back of the head. In less than a minute one boy fell into a sound sleep, from which, however, he was easily awakened; the other, though the magnetising process was continued for upwards of a quarter of an hour, seemed to experience no effect of any sort. He appeared, however, marvellously to enjoy his position; and more than once, as though to tantalize the curious group around, he feigned to sleep for a moment, and then, suddenly opening his eyes, laughed in the face of those who were anxiously bending forward to witness the unexpected success which they imagined was crowning the experiment. The sleep of his companion seemed natural and unfeigned, and after having been wakened, he was again thrown into a state of slumber by a repetition of the process. It was stated, that the boy who had been magnetised had not experienced a single fit since the first operation, while the other, upon whom no effect had been produced, though treated in the ordinary and approved method, had suffered from them every day.

After the boys had been dismissed, a girl, apparently of about fourteen years old, was brought forward. She was operated upon in the same manner as had been adopted in the former case. It was about three minutes before any perceptible result was produced; at the expiration of which period she was in a state of profound and obviously unconscious slumber. It should be stated, that she had been brought into the hospital on account of the loss of all use of the lower extremities, produced by an hysterical affection. She had been treated in the ordinary way without success; but subsequently to her being magnetised, she had recovered the use of her limbs, and walked unsupported into the room. Two or three attempts were made to rouse her by lifting her arm, pulling her hair, calling her, lifting her off the chair, and placing her upright, but without the least effect. She was then half awakened, or rather thrown into a state of somnambulism, by the operator drawing his thumbs across the eyebrow, from the insertion of the nose outwards; and in this condition, though still seemingly unconscious, she was able to stand upright. After being kept in this state for a short time, she was again thrown into a sound sleep, and placed in a chair, while the next subject was introduced. In this chair she slept for about an hour and a half, through a hubbub that might have awakened the seven sleepers.

Her successor was a girl of about the same age, of quiet, demure, almost sullen countenance, subject to occasional fits of delirium. She was speedily thrown into a deep sleep, and was then somnabulised. When in this state of partial consciousness—a consciousness like that of true somnambulism, when the individual is fully aware of what he does, though all memory of it ceases with the cessation of his sleep—the whole expression of her countenance changed. It became cheerful, animated, and intelligent. She noticed objects immediately in front, though of the presence of all but these she seemed absolutely ignorant; and she made some not very complimentary observations upon the appearance, dress, and manners of some of the persons present. Many experiments were tried upon her, of the most of which it was scarcely possible that she should have been conscious, for the purpose of